Genetically Engineered Organisms: Environmental and Economic Risk Assessment

Genetically Engineered Organisms: Environmental and Economic Risk Assessment

Lexie Reid

Callisto Reference

www.callistoreference.com

Callisto Reference,
118-35 Queens Blvd., Suite 400,
Forest Hills, NY 11375, USA

Visit us on the World Wide Web at:
www.callistoreference.com

ISBN: 978-1-64116-790-1 (Hardback)

Cataloging-in-Publication Data

Genetically engineered organisms : environmental and economic
risk assessment / Lexie Reid.
p. cm.
Includes bibliographical references and index.
ISBN 978-1-64116-790-1
1. Transgenic organism. 2. Genetic engineering. 3. Organisms. I. Reid, Lexie.
QH442.6 .G46 2023
636.082 1--dc23

Table of Contents

Preface

Every book is initially just a concept; it takes months of research and hard work to give it the final shape in which the readers receive it. In its early stages, this book also went through rigorous reviewing. The notable contributions made by experts from across the globe were first molded into patterned chapters and then arranged in a sensibly sequential manner to bring out the best results.

Genetically engineered organisms or genetically modified organisms (GMO) are animals, plants or microorganisms whose DNA has been altered using genetic engineering techniques. They can be classified into categories such as genetically modified crops, genetically modified medicines and genetically modified animals. Genetically modified crops are the crops whose genes have been modified. They are also called transgenic crops. The main objective of developing these crops is to increase the yield and to produce disease-resistant crops. These crops have entered the markets and have become a part of the global food supply chain. However, genetically modified animals are mainly restricted for research purposes. This book outlines the environmental and economic risks and benefits associated with genetically engineered organisms. Some of the diverse topics covered herein address the varied organisms that fall under this category. The book is appropriate for students seeking detailed information in this area as well as for experts.

It has been my immense pleasure to be a part of this project and to contribute my years of learning in such a meaningful form. I would like to take this opportunity to thank all the people who have been associated with the completion of this book at any step.

Lexie Reid

1

SPAGE Techniques and Technology

Johannes L. Frieß, Bernd Giese and Arnim von Gleich

Introduction

In recent years, innovation in genetic engineering brought forth a number of technologies to manipulate the fate of entire wild type populations. These technologies rely on the dissemination of synthetic genetic elements within a population of sexually reproducing species via the germline and are identified as Self-Propagating Artificial Genetic Elements (SPAGE). Some secure their dissemination passively so that only offspring carrying the SPAGE will survive or be fertile. Others overcome the limitations of the Mendelian inheritance pattern by a distortion of allelic segregation or a fragmentation of chromosomes, resulting in e.g. an altered sex ratio. Genetic elements may also promote their preferred inheritance by a molecular mechanism. If a SPAGE overcomes the Mendelian pattern of inheritance and is thereby enabled to spread and distribute a novel trait throughout a population – even defying natural selection – it is called a gene drive. If organisms have a comparably short generation time, as e.g. insects, then already after a few months, a large part of the population could express a new property transmitted by the gene drive. In particular, very invasive gene drives may be able to impose properties on entire populations that otherwise could not spread.

SPAGEs are discussed for many potential applications and partially already designed as a kind of self-propagating delete-function. If for example the property mediated by the synthetic genetic element consists in male or female offspring becoming infertile, an entire population may disappear.

J. L. Frieß (✉) · B. Giese
Institute of Safety/Security and Risk Sciences, University of Natural Resources and Life Sciences, Vienna (BOKU), Vienna, Austria
e-mail: Johannes.friess@boku.ac.at

A. von Gleich
Department of Technological Design and Development, Faculty Production Engineering, University of Bremen, Bremen, Germany

Currently, multiple applications are under consideration. Especially malaria- or dengue-carrying mosquitoes are potential targets. In agriculture, weeds and crop pests could be eradicated or endangered species could be immunized against pathogens using a GD. Two potential applications of gene drives even serve issues of nature conservation, namely the eradication of invasive animal or plant species (Webber et al. 2015) and the conservation of endangered species (Esvelt et al. 2014; European Commission and Scientific Advice Mechanism 2017; Ledford 2015). Although discussed in the 2016 NASEM report on gene drives (National Academies of Sciences 2016), the idea to recover the sensitivity of pest species to pesticides or to remove transgenic resistances from feral populations have not been pursued in the scientific literature of the following years. So far, gene drives have not yet been released, but the discussion is gaining momentum (Courtier-Orgogozo et al. 2017; Emerson et al. 2017; Hochkirch et al. 2017). In particular, the development of new gene drive variants is closely linked to the upswing that genome editing methods have taken by the recent use of CRISPR-Cas gene scissors (Gantz et al. 2015; Gantz and Bier 2015).

Compared to previous releases of GMOs, SPAGE and especially gene drives collide with basic principles of precaution due to their targeted property to spread in wild populations and thus causing extreme exposure. Applications of this new quality represent a shift of paradigm in the handling of GMOs. At least for the European Community, the current regulation of the release of GMOs assumes that for specific periods of time a certain amount of GMOs will be released in a particular region.[1] However, now a type of genetic technology arises whose innermost principle lies in exceeding these limits: the transformation or even eradication of wild populations.

So far, it is unclear whether particular SPAGE applications, once released, will be retrievable or manageable at all. Due to their intended ability to spread, a loss of control is highly probable, not least in comparison to hitherto existing GMOs. In general, SPAGEs must be characterized as a technology with a high depth of intervention into the genetic configuration of organisms and ecological systems, which results in a high technological power (much higher compared to a manipulation at the phenotype level and with high potential impacts with regard to the functionalities of the modified organisms) and a high range of exposure (spread in space and time because of self-reproduction, mobility and self-dispersal). Due to the increased ability to self-propagate and spread through populations, a particularly high exposure of these altered organisms to ecosystems must be expected. This dispersion and exposure must in turn also be appraised as a high depth of intervention into the targeted ecosystems, which additionally may be regarded as a contamination of these systems. The increased technological power and exposure produced by these technologies results in proportionally increasing lack of knowledge about possible consequences, reaching from enormous scientific uncertainties to vast ignorance. A correspondingly extended risk assessment (hazard and exposure assessment) is required to fathom the extent and the depth of these dimensions of non-knowledge

[1]Cp. Annex III A and Annex III B of the EC Directive 2001/18/EC of the European Parliament and of the Council on the deliberate release into the environment of genetically modified organisms (EC Directive 2001/18).

on the different organizational levels of biosystems that are affected by gene drives. The question arises whether methods and models are already available to adequately investigate hazard and exposure potentials caused by such a wide spread of new properties in whole populations and possibly into related species. Above all, the effects of a strong reduction of populations up to their eradication are important and complex evolutionary changes that must be considered. Important questions regarding technological, ecological as well as ethical issues become apparent.

Besides a review of modes of action of current SPAGE-technologies, the following technology characterisation analyses the depth and intensity of intervention, the resulting potential power and exposure and the corresponding extent of non-knowledge associated with the different SPAGE-techniques in a comparative approach.[2]

SPAGE-Techniques

A variety of different self-propagating artificial genetic elements has been developed in recent years. Some of them occur naturally as selfish genes and have been further optimized, while others are genetically engineered.

The SPAGEs currently in development or already applied encompass:

- RIDL-technology ("release of insects carrying a dominant lethal genetic system")
- Meiotic Drives (autosomal- or Y-linked X-shredder)
- Killer-Rescue
- Maternal Effector Dominant Embryonic Arrest (Medea)
- Underdominance-based systems
- Homing Endonuclease Genes (HEG) based systems, especially CRISPR/Cas.

SPAGEs are developed to manipulate genes and traits of organisms and thereby alter whole populations to serve particular needs. Meanwhile a panoply of applications is envisaged for SPAGE. For instance, they should be applied to fight infectious diseases in particular the vectors of diseases. Especially malaria- or dengue-carrying mosquitoes are potential targets. In agriculture, weeds and crop pests should be eradicated or endangered species could be immunized against pathogens using a gene drive. Two potential applications of gene drives even serve issues of nature conservation, namely the eradication of invasive species (Webber et al. 2015) and the conservation of endangered species (Esvelt et al. 2014; Ledford 2015; Champer et al. 2017; European Commission and Scientific Advice Mechanism 2017).

Once released into the environment, SPAGEs can hardly be retrieved. Although there are many ideas to restrict their spread or even to alleviate adverse effects,

[2]The technology characterisation was in large parts already published.

Frieß JL, von Gleich A, Giese B, 2019. Gene drives as a new quality in GMO releases - a comparative technology characterization. PeerJ 7:e6793.

a complete reversal and restoration of the pre-existing state (and genotype) seems impossible.

So-called self-limiting approaches may pose a partial exception to this as their mode of action is developed to result in a successive frequency decrease of the SPAGE within a population.

Gene Drives

A number of natural mechanisms possess this notable property. In 2006, Sinkins and Gould mentioned transposable elements, meiotic drive genes, homing endonuclease genes and *Wolbachia* as naturally occurring gene drives. A theoretical concept for gene drives as a method to drive a desired gene, or a set of genes into a population was already proposed in 1960 by Craig et al.: "Mass release of male-producing males might be used in control operations" (Craig et al. 1960). The spread of chromosomal translocations was already proposed as a means of population control (Curtis 1968; Serebrovskii 1940). Hastings suggested to use so called "selfish genes" for that purpose (Hastings 1994) and a practical implementation was explored with the use of the P-element for germline transformation of *Drosophila melanogaster* (Carareto et al. 1997). Austin Burt in 2003 suggested to use homing endonucleases for the design of self-replicating drives (Burt 2003). Gene drives propagate even if they confer a fitness penalty, or in other words "Mathematically, drives are initially favoured by selection [...] if the inheritance bias of the drive exceeds its fitness penalty." (Noble et al. 2018, p. 201). Some secure their dissemination passively so that only offspring carrying genetic information of the drive will survive or be fertile. Akbari et al. (2015) called this type of mechanism "selective embryonic lethality". Others actively overcome the limitations of the Mendelian inheritance pattern by a distortion of allelic segregation i.e. fragmentation of chromosomes, for example resulting in an altered sex ratio. Active drives may also copy their genetic information between homologous chromosomes resulting in homozygous offspring. Such approaches were termed „active genetics " by Gantz and Bier (2015). In the sense of such a broad definition most of the SPAGE-technologies mentioned above can be regarded as gene drives. Due to the exclusion of *Wolbachia*-based techniques in this work, only certain RIDL-approaches with a self-limiting character represent exceptions.

Due to its inherently 'invasive' character, a once-released gene drive represents a significant intervention into ecosystems. In principle, a gene drive needs several generations to establish itself in a population. It is thus a technology capable to reproduce itself and undergo mutational changes over time. Not only do gene drives affect the environment, the environment affects the gene drives as well. A gene drive engineered in the laboratory, once released will be confronted with evolutionary processes.

Methodology of Technology Characterisation

Technology characterisation is an approach for prospective technology assessment that is applicable extremely early in the innovation process, when results of scientific research and the outlines of the technology are quite well known, but possible applications and affected systems are still unclear. This actually is the case with SPAGE-technologies. Such an early assessment is important and useful, because in case of severe concerns mitigations, corrections and course changes to alternative development paths are more easily directed and much more cost-efficient before large investments into products and production facilities are made. Technology characterisation is in this way an important approach to operationalize the requirements of precaution. The aim of technology characterisation is the early assessment of potential hazards and exposures (identifying reasons of concern) and, possibly still more important, the assessment of different dimensions and forms of lacking knowledge regarding hazards and exposure, reaching from uncertainties to absolute ignorance. This is the only way to include complete surprises, which means possible events for which currently no scientific approved 'model of effect' exists.[3] However, approaches for an early assessment of potential hazards and exposure as well as an assessment of different dimensions and forms of lacking knowledge regarding hazards and exposure already exist (Ahrens et al. 2005; Giese and von Gleich 2015; Linkov et al. 2018; Owen et al. 2009; Steinfeldt et al. 2007). The underlying hypothesis of technology characterisation is, that the range and the forms of non-knowledge are not 'just there', but are to a large extend produced by the characteristics of the technology. Depth of technological intervention and also the intensity of intervention are the first criteria to investigate the range and forms of lacking knowledge (from uncertainties to ignorance) by scrutinizing their technological origin. The depth of intervention is a source of enormous technological power and therefore of mighty potential effects, benefits as well as hazards, on one side. On the other side, the depth of intervention presents sources of a high operating range of the created entities and thus the potential for exposure. High power and high range of exposure lead to a high extend of non-knowledge concerning possible effects and interactions. In order to provide additional information on the frequency and the corrigibility of the expected effects, the quantitative aspects of the use of the technology (intensity of intervention i.e. quantity, frequency of its application), its reliability in practice, the probability of failure, and, finally, possible ways of limiting harm in case of failure have to be analyzed.

The aim of prospective technology characterisation is not to identify any possible adverse effect of technologies. Most of the occurring adverse effects will also in future be manageable by trial and error. Instead, it should provide a basis for decision-making in the view of the precautionary principle (Commision of the European

---

[3] As it was the case with DDT minimizing the thickness of bird eggs, ozone depletion triggered by CFC, the 'mad cow disease' and industrial chemicals functioning as endocrine disrupters (European Environment Agency 2002) and is actually the case with the reduction of insect populations in Middle Europe.

Communities 2000; The Rio declaration on environment and development 1992; United Nations 2000). "The precautionary principle enables decision-makers to adopt precautionary measures when scientific evidence about an environmental or human health hazard is uncertain and the stakes are high" (European Parliament Think Tank 2015) The precautionary principle legitimates precautionary action in cases when it is unwarrantable to wait until a risk is clear and proven, because a probably occurring disaster will then not be controllable. Preconditions for precautionary action are therefore: (a) lack of knowledge (from uncertainty to ignorance), (b) comprehensible reasons for concern (affecting extremely powerful and/or far reaching consequences), (c) a rudimentary cost–benefit analysis (in which e.g. medical applications with few less risky options are rated higher than applications in the food chain with plenty alternatives), (d) adequate precautionary measures (reaching from containment over substitution by less problematic alternatives to moratorium (Fischer et al. 2006). In our approach to operationalize the precautionary principle, the focus of technology characterisation lies on the prevention of far reaching, by-trend irreversible and global effects of events with adverse consequences that cannot be managed adequately, that cannot be retrieved, corrected or mitigated in case of their occurrence.

Based on the framework for technology characterisation SPAGE-technologies will be compared considering the following criteria (also depicted in Fig. 1.1).

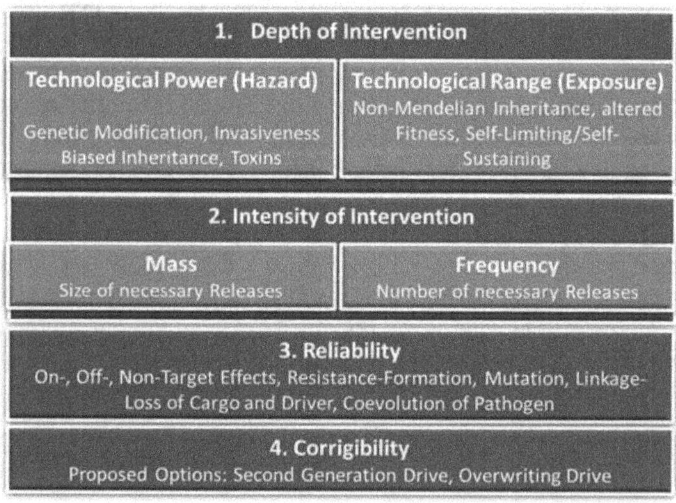

Fig. 1.1 Criteria of prospective technology characterisation with corresponding gene drive-specific effects and options. Technological power is not in the focus of this rather general prospective study due to the very early innovation phase, where the particular design (esp. their cargo) and application context of gene drives is not yet clear

Depth of Intervention (Technological Power and Range)

Depth of intervention results in high technological power and range. For SPAGEs in general, the depth of intervention is much higher than in approaches for breeding and population control which are not based on genetic modifications. SPAGEs like other genetic engineering technologies constitute a manipulation of the very basis of organisms, their genetic characteristics, which results in one source of their technological power. The other source of power lies in the functionalities of the applied genes resp. traits. The technological range describes the potential spatio-temporal consequences of a gene drive, considering its lasting persistence in a population as well as the range with which it could spread by mobility and across populations. But there are differences. The mono-generational suppression of a single population is considered as a comparably low range, while the permanent replacement of a population with genetically altered specimens is considered a high range. At the same time, range considers the possibility of either intended or unintended spread of a gene drive across multiple populations (invasiveness).

Intensity of Intervention (Number/Frequency)

The intensity of intervention as number and frequency of released organisms describes the necessary quantity of interventions to drive a desired trait into a targeted population. An approach requiring the released organisms to outnumber the wild type organisms would score as high intensity and if an initially low percentage of the population is sufficient it would correspond to a comparably lower intensity of intervention. The quality of released organisms, e. g. their capability of self-reproduction, which determines their range in a much higher proportion is determined by the criterion of depth of intervention. In terms of intensity, without changing the depth of intervention, power and range of a technology are only dependent on number and frequency of its application.

Reliability of the Technology

Reliability describes the probability of failure of the technology with regard to its intended use (unintended side effects and long term effects). Important reliability issues are e. g. linkage-loss of the cargo gene and its driver system, the generation of resistances in the target population, coevolution of the pathogen and system decay (Alphey 2014).

Corrigibility or Limitation of Damage in Case of Failure

This criterion addresses an important aspect of risk management. Can a gene drive be retrieved in case that something goes dramatically wrong and can the damage of a failed gene drive be reversed by any means and if so, how laboriously are they compared to the initially released construct/system? For some SPAGE technologies it is claimed that they can be somewhat remedied by a release of wild type organisms. But such an endeavour would not really reverse the damage done. Even more difficult to estimate are corrective actions such as a reversal drive which on one hand relies on the release of a second generation gene drive to remedy the failures of the first. And on the other hand, the gene pool of the target population in any case retains transgenic elements.

Important Preliminary Remarks

As a means to compare the different SPAGE technologies, a rough assessment into three classes as high, moderate and low was applied, based on the criteria introduced above. Regardless of the fact that such a classification has to be further differentiated in subsequent studies, it has to be noted that this rating only refers to the comparative approach between the technologies included in this study and cannot be used to draw any conclusion on their absolute impact, for instance in comparison to solutions avoiding genetic technologies.

SPAGE Technologies

In the following, most of the SPAGEs taken into consideration to be used in a gene drive are discussed, examining their methodological advantages and drawbacks. Furthermore, they are compared and analysed in a hypothetical application concerning the aforementioned criteria of technology characterisation, which are further differentiated in the subsequent paragraph.

Other molecular components, such as transposable elements, TALENs (transcription activator-like effector nuclease) and ZFNs (zinc finger nuclease), which in theory could be considered as being potentially involved in gene drive mechanisms were not included into this technology characterisation due to their low transformation rate, limited specificity or problems to deal with sequence polymorphisms in wild populations.

Release of Insects Carrying a Dominant Lethal Gene (RIDL)

In this approach, laboratory-reared organisms (until now flies, mosquitoes and a moth species), equipped with a dominant lethal gene, are mass released to reduce the number of offspring in a wild population. There are two varieties of RIDL. In the bi-sex RIDL approach, the offspring of both sexes die in the zygotic, larval or early pupal stage. In the female specific RIDL approach (fsRIDL), only female offspring die. Male offspring is heterozygous for the dominant lethal gene and therefore would pass on the lethal trait to 50% of their offspring in subsequent generations. Female specific RIDL strains have been developed for *Aedes aegypti* and *Aedes albopictus*, using flightlessness as a lethal trait (Alphey et al. 2013). Since RIDL as well as fsRIDL organisms will be selected from the population within a few generations, the RIDL technology can be considered self-limiting.

Depth of RIDL Intervention

The power (by quality) of the bi-sex RIDL technology is based on the functionality of the applied gene comparably high due to its capacity of killing 100% of its offspring but its range can be seen as low in comparison to other SPAGE techniques, being self-limiting, it is not intended to persist in the ecosystem.

Although only half of the GMO's offspring survives, fsRIDL scores as a high power and low range technology due to its self-limiting and suppressive quality with the potential to eradicate the population with a concomitant risk of invasion into neighbouring populations, regarding its high release ratios.

Intensity of RIDL Intervention

Independent of the applied RIDL variant the technology requires mass releases of genetically modified organisms. In the case of the bi-sex RIDL even a higher number of mass releases may be necessary. Oxitec has used release ratios of up to 54: 1, before it observed a reduction in wild populations (Gene Watch UK 2013). The mass and frequency for a successful application can therefore be rated as high.

Reliability of the bi-Sex RIDL Technology

The overall probability of failure of the bi-sex RIDL technology compared to other SPAGE techniques is moderate. This evaluation strongly stems from the issues arising from the first field test trials. Key points of error encompass:

- lowered fitness of laboratory-reared GM insects due to inbreeding (colony effect)
- reduced mating capabilities of the GMOs (reducing the suppressive effect)
- selection against the fitness burden which the dominant lethal gene clearly poses,

- the genetic bi-stable switch necessary to rear the flies in the laboratory requires tetracycline (if tetracycline is present in the target ecosystem, the suppressive effect would be reduced and protracted),
- errors in the release:
  - wild types (would reduce the suppressive effect), and
  - phenotypic wild types carrying the non-functioning dominant lethal gene (would reduce the suppressive effect and persistently introduce synthetic DNA into the ecosystem).

Potential Vulnerabilities of the Target System Towards the bi-Sex RIDL Technology

- adverse ecological effects due to the eradication of the target population due to the ratio of released GMOs to the native population,
- spread to other populations,
- toxicity of the dominant lethal gene product (tTA) to predators,
- dead GM-larvae are likely to enter the commercial food production and food chain.

Reliability of the fsRIDL Technology

The overall probability of failure specific for the fsRIDL technology is moderate compared to other SPAGE techniques. Most of the aforementioned weaknesses within the technology however, in practice are alleviated by increased release numbers. Nevertheless, key points of error encompass:

- The selection against the fitness burden of the dominant lethal gene which seems more likely in a system with longer persistence, yet implausible within the few generations fsRIDL organisms persist.
- errors in the release:
  - wild types (would reduce the suppressive effect),
  - female fsRIDL-organisms (would further enhance the suppressive effect in the first generation), and
  - phenotypic wild types carrying the non-functioning dominant lethal gene (would reduce the suppressive effect and persistently introduce genetically modified DNA sequences into the ecosystem).

Potential Vulnerabilities of the Target System Towards the fsRIDL Technology

- adverse ecological effects due to the eradication of the target population due to the ratio of released GMOs to the native population,
- spread to other populations (which may be especially probable in a population consisting mainly of fsRIDL males with a dwindling percentage of females),
- the genetic bistable switch necessary to rear the organisms in the laboratory requires tetracycline (if tetracycline is present in the target ecosystem, the suppressive effect would be reduced and protracted),
- toxicity of the dominant lethal gene product (tTA) to predators.

Possibilities for Limitation of Damage Caused by RIDL Technology

Independent of the applied RIDL variant, the only feasible option for an attempt to restore the original population after detrimental effects caused by RIDL technology is the release of wild type specimens. However, since bi-sex RIDL is self-limiting and if fsRIDL-organisms will be selected from a population within a number of generations, the damage of a failed release might be buffered by the resilience of most of the ecological systems as long as it does not result in a complete eradication of the target species or other parts of the food web.

Meiotic Drives (MD) in particular X-Shredder

Meiotic Drives (MD) consist of selfish genetic elements which cause a distortion of allelic segregation that results in a bias of the frequency of Mendelian inheritance. For instance, the Mendelian segregation frequency of 50% is distorted up to 70% in *Zea mays* (Australian Academy of Science 2017; Lindholm et al. 2016). Other MDs have been reported for *Drosophila melanogaster* (segregation distorter [SD] system) (Larracuente and Presgraves 2012), the mouse *Mus musculus* (t-haplotypes, causing a transmission ratio distortion) (Silver 1993), *Zea mays* (abnormal chromosome 10 [Ab10]), which affects Gonotaxis, distorted sex ratios in *Silene* species (Taylor 1994) and mosquitoes. In the latter, MDs are naturally occurring in *Aedes aegypti* (Craig et al. 1960) and *Culex pipiens* (Sweeny and Barr 1978). A major drawback of MDs consists in the fact that the fitness of other alleles at the same locus, which do not bias transmission, and alleles linked to them, is reduced (Lindholm et al. 2016).

For gene drives, a particularly interesting MD is the so called X-Shredder, which causes fragmentation of the X chromosome by nucleases during male meiosis. Thereby, only Y-bearing sperm can produce viable offspring, which is of course male (Newton et al. 1976). An autosomal X-shredder can be regarded as self-limiting, a Y-linked X-shredder as self-sustaining (Burt 2003; Burt and Trivers 2006; Deredec et al. 2008). A Y-linked X-shredder can invade adjacent populations or species with incomplete mating barriers, therefore widespread effects may be anticipated (Alphey 2014). Galizi et al. (2014) published a synthetically engineered X-shredder aiming at spermatocyte meiosis in *Anopheles gambiae*, producing mainly Y-chromosome-carrying sperm, causing a male bias of up to 95%. A distortion of the sex ratio is a penalty to fitness, which may in extreme cases lead to a population's extinction. Although rarely, in Drosophila, sometimes 100% female progeny is achieved. Therefore, this trait is highly selected against. Hence, meiotic drive-based extinction has never been observed in natural populations (Helleu et al. 2015). For this study we focus on the self-sustaining variant of the Y-linked X-Shredder.

Depth of X-Shredder Intervention

The X-Shredder approach, considering a male bias up to 95% (Galizi et al. 2014), would cause a major population suppression, therefore its technological power and range are rated as high. Since it constitutes a self-sustaining gene drive and since a population consisting mainly of males is much more likely to migrate in search of females its range is also considered as high. Although, to date no field trials with X-shredder gene drives have been undertaken, and therefore no migration patterns of X-shredder males are available.

Intensity of X-Shredder Intervention

The X-Shredder approach requires a mass release of males. The necessary technological input generating power and the range (by quantity) can thus be regarded as high. However, even a small release size would theoretically suffice to replace a population over multiple generations, dependent on the fitness of the gene drive organisms.

Reliability of the X-Shredder Technology

Based on the small number of available publications on X-Shredder approaches in a preliminary comparative assessment of SPAGE technologies, the failure probability is not easily estimated but it can comparatively be considered as moderate. Key points of error encompass:

- selection against the fitness burden which the construct clearly poses,
- errors in the release:
  - *phenotypic male wild types carrying the non-functioning construct (would reduce the suppressive effect).*

Vulnerability of the Target System Towards the X-Shredder Technology

- adverse ecological effects due to the eradication of the target species (although not yet being observed in nature) and
- spread to other populations (which seems likely in a population with a dwindling number of females).

Possibilities for Limitation of Damage Caused by X-Shredder Technology

There is no possibility to directly remedy the damages obtained from an X-Shredder release. This makes the technique highly problematic, it is built to first invade and replace followed by immediate eradication, due to the lack of females. Its low threshold quality further exacerbates the handling of Y-linked X-Shredder gene drives.

Killer-Rescue

The Killer-Rescue System was first proposed by Gould et al. (2008). It consists of two unlinked loci one encoding a toxin (killer allele), the other encodes an antidote (rescue allele) (Gould et al. 2008). Thereby, the toxin and antidote could consist of miRNAs and recoded gene or a toxic protein and detoxicating enzyme. Furthermore, a cargo gene can be fused to the antidote gene. Homozygous carriers of both genes would be mass-released into wild populations, offspring which inherit the killer allele but not the rescue allele would be non-viable. Since both alleles are not linked in their inheritance the killer allele will be quickly selected from the population, while the rescue allele confers a clear fitness gain and will increase in its prevalence. As soon as the killer allele has completely disappeared from the population however, the fitness gain of the Rescue allele will disappear as well, unless the cargo gene confers a gain in fitness. This system is designed to be a self-limiting modification gene drive in which, if the cargo gene bears a fitness penalty, its prevalence in the population would decrease after a number of generations. There is a possible variant where multiple copies of the killer allele are incorporated into the GDOs' genome, enhancing the selective benefit of the rescue allele. A particular benefit of the technique is that it is easy to design and engineer.

Depth of Killer-Rescue Intervention

The Killer-Rescue system's technological power compared to other SPAGE techniques scores as low because the killer-allele will potentially cause only a short-term reduction in the population size, it is not by design a suppression drive. Considering the technological range, Killer Rescue, due to its non-persistent quality and its therefore limited chance of contamination of other populations and relative high invasive threshold scores as low.

Intensity of Killer-Rescue Intervention

The Killer-Rescue system is reliant on a high number of released carriers of up to a ratio of gene drive organisms to wild types of 2:1, according to model scenarios by (Gould et al. 2008). Although this ratio is much lower than reported for other mass release dependent SPAGE techniques, the mass and frequency of Killer-Rescue still has to be scored as high.

Reliability of the Killer-Rescue Technology

The Killer-Rescue system's probability of failure scores as low in comparison to other SPAGE-techniques. Although there is no data on most of the imaginable vulnerabilities of the technology. It would be recommendable to use miRNA as a killer allele in order not to give the carrier-organisms a toxic quality. Key errors encompass:

- the selection against the fitness burden which the constructs clearly pose (resistance formation or toxin-inactivation).

Vulnerability of the Target System Towards the Killer Rescue Technology

Imaginable key vulnerabilities of the Killer-Rescue system encompass:

- Linkage loss
- Natural evolution of an antidote
- Inactivation of the killer allele.

Possibilities for Limitation of Damage Caused by Killer-Rescue Technology

Since it is expected that the Killer-Rescue system has a high invasion threshold (although lower than that of two-locus Underdominance), the most feasible option to limit the spread of this gene drive is a release of wild types.

Maternal-Effect Dominant Embryonic Arrest (Medea)

The term Medea is an acronym named after the sorceress in Greek mythology who killed her own children. This name is accurate as a Medea selfish genetic element consists of two chromosomally-located tightly linked transgenes: one that encodes a (miRNA-)toxin inherited by all progeny of Medea-bearing mothers, and a second that encodes an antidote (gene without miRNA-sequence) active in the zygote (Akbari et al. 2014). Therefore, only Medea-bearing offspring (hetero- or homozygous) survives. This maternally induced lethality of wild type offspring not inheriting a Medea allele grants an ability to invade populations.

The Medea elements were first discovered in *Tribolium* flour beetles and have also been reported in mice. The only published Medea constructs (*Medea myd*88, o-fut1 and dah) have been inserted on an autosomal chromosome in *D. melanogaster. myd88* is a maternally expressed gene required for embryonic dorso-ventral pattern formation.

If *Medea* is located on the X chromosome in a X/Y male heterogametic species, *Medea* is predicted to spread to allele fixation, with wild type alleles being completely eliminated (Akbari et al. 2014).

Medea organisms exhibit a high-frequency stable equilibrium when the transgenic construct is associated without any fitness cost (Gokhale et al. 2014). The fitness costs of homozygote Medea Drosophila were estimated to be 27.3% and 17.4%, respectively, for two different targeted genes. In lab trials, where 25% of the original members were homozygous for *Medea*, the gene spread through the entire population within 10 to 12 generations. Observations indicate that a single copy of each *Medea* toxin is sufficient to induce 100% maternal-effect lethality and a single copy of each rescue transgene is sufficient to rescue normal development of embryos derived from mothers expressing one or two copies of the toxin (Akbari et al. 2014). Until now, attempts to establish a Medea system for *Aedes aegypti* were not successful. There are other variants of single locus constructs, such as Semele and inverse Medea. Semele confers toxic sperm that either renders females infertile or kills them. In inverse Medea the promotors of the toxin and the antidote are switched (Marshall and Akbari 2015).

Currently Medea is planned to be applied in order to take control of the cherry fruit fly (spotted-wing fruit fly *Drosophila suzukii*) in California (Regalado 2017). Two considered approaches are to either target female fertility genes or to alter the ovipositor of the flies to make them unable to puncture the ripening cherries. Buchman et al. (2018) found pre-existing native resistances against the miRNA toxins of their construct in 5 out of 8 examined *D. suzukii* strains. Together with the high fitness penalties conferred by the construct, the Medea GD now has to be considered a high threshold drive, that, when conferring a large fitness penalty, will only be transiently maintained in the population without supplemental releases. In a mathematical model for the *myd88* construct in the cherry fruit fly, a fitness cost for heterozygotes of 28 and 65% for homozygotes was assumed (Buchman et al. 2018).

Depth of Medea Intervention

In a comparative approach of SPAGE the technological power of Medea is to be rated as moderate, as it will certainly drive to fixation and has therefore only a potentially transient effect on population size. The range and thus its potential of exposure and contamination would score high due to its higher invasiveness and potential to invade non-target populations.

Intensity of Medea Intervention

In theory, it would not require many carrier organisms to drive a gene into a population. Therefore, the intensity of intervention would have to be rated as low due to its low number and frequency required for a successful approach, compared to other SPAGE techniques. However, as demonstrated by Buchman et al. (2018), due to the pre-existing resistances and high fitness penalties it is more likely that multiple mass releases are required for a successful drive (Marshall et al. 2017). Therefore, a high intensity of intervention is considered for Medea gene drives.

Reliability of the Medea Technology

The Medea technology's probability of failure scores as low in comparison to other SPAGE-techniques. This is founded on its low probability of linkage loss, resistance formation, and its potent toxin- and rescue-mechanism. Thereby, the technique does not rely on toxins that might harm other organisms upon ingestion but on RNAs which degrade quickly outside the cells. Key points of error encompass:

- the selection against the fitness burden which the constructs clearly pose (resistance formation by toxin-inactivation)
- errors in the release:
  - Medea-males, homozygous Medea-females (would protract the suppressive effect and accelerate the genes' drive to fixation),
  - wild types (would reduce and protract the suppressive effect).

Vulnerability of the Target System Towards the Medea Technology

- adverse ecological effects due to the permanent introduction of engineered genes into the ecosphere and their effects on population dynamics and
- effects due to the spread to other populations (will almost certainly happen over time).

Possibilities for Limitation of Damage Caused by Medea Technology

A potential measure would be to release a second generation Medea gene drive. This would introduce a new toxin-antidote combination as well as the antidote for the first generation toxin. Although the suppressive effect of Medea may be stopped by this approach it introduces even more persisting GMOs in the ecosystem.

Underdominance (UD)

Underdominance, also known as heterozygote inferiority, is a genetically engineered gene drive technique. There are two different approaches UD^{mel} (Akbari et al. 2013) and *Rpl14* (Reeves et al. 2014). One approach is operated by two gene constructs. Each construct consists of a maternal toxin gene and an embryonic antidote. However, the antidote to each toxin is located on the other construct. Thus, an embryo needs both constructs in order to have both antidotes to the maternally administered toxins. Therefore, UD heterozygotes have a lower fitness than homozygotes (Reeves et al. 2014). The constructs can be located on the same chromosome or on different chromosomes (two-locus Underdominance). When a UD female heterozygous for both constructs mates with a wild type male, 25% will be heterozygous for both constructs, while 25% of offspring will be non-viable wild types, and 50% will be non-viable due to the lack of one of the necessary antidotes. The toxins of Underdominance constructs may be the same as utilised in the Medea technology: *myd88*, *dah* and *o-fut-1* (Akbari et al. 2013) or *RpL14*, a cytoplasmic ribosomal protein which is haploinsufficient (Reeves et al. 2014). Since these toxins are administered maternally, a release of wild type males into a replaced Underdominance population would lead to a population crash, as all offspring would inherit the wrong antidote (Akbari et al. 2013). A UD gene drive requires a high threshold release (National Academies of Sciences 2016). For the *RpL14* construct, this threshold is estimated to be as high as 61% of the total population (Reeves et al. 2014).

Therefore, an intentional underdominant population transformation is inherently reversible where it is realistically possible to release sufficient wild type individuals to traverse the unstable equilibrium in the lower frequency direction (Gokhale et al. 2014).

Depth of Underdominance Intervention

The power of the Underdominance approach has to be rated as moderate compared to other SPAGE techniques. This effect will at first persist but eventually fade over the subsequent generations. In comparison to the Medea approach, the range of Underdominance is estimated to be lower, due to its higher invasion threshold.

Intensity of Underdominance Intervention

Since utilisation of this technology is more frequency-dependent than the Medea approach, requiring even greater mass releases, its intensity generating quantitatively mass and frequency is rated as high.

Reliability of the Underdominance Technology

The overall failure probability of the Underdominance technology in comparison to other SPAGE techniques can be estimated as moderate. Key points of error encompass:

- lowered fitness of laboratory-reared GM insects due to inbreeding (colony effect)
- the selection against the fitness burden which the constructs clearly pose.

Vulnerability of the Target System Towards the Underdominance Technology

- adverse ecological effects due to the eradication of the population due to its small size and
- spread to other populations.

Possibilities for Limitation of Damage Caused by Underdominance Technology

For a UD drive the release of wild type specimen represents the most obvious option to potentially restore the original population.

Homing Endonuclease Genes (HEG)

HEGs are selfish genetic elements. But different from transposable elements, they code for a restriction enzyme with a target sequence of 20–30 bp. The HEG is nestled within its own recognition site. An expressed homing endonuclease-protein finds intact recognition sites and cuts them. Then the selfish genetic element relies on the DNA-repair mechanism of homologous recombination which copies the HEGs code and inserts it into the cut-site on the homologous chromosome.

CRISPR/Cas9

CRISPR stands for **C**lustered **R**egularly **I**nterspaced **S**hort **P**alindromic **R**epeats, while Cas stands for **C**RISPR-**as**sociated protein. Both components originate from an adaptive immune system of bacteria and archaea. Cas9 is a ribonucleoprotein (RNP), able to bind guide RNAs (gRNA), aka crRNA that specifically recognize and bind to the target sequences (20 nucleotides). The target DNA-sequence must contain a protospacer adjacent motif (PAM) with the sequence NGG (N can be any nucleotide) for the Cas protein to cut. The cut takes place three nucleotides upstream of the PAM. The Cas protein can cut at multiple PAMs as long as they are at least 8 nucleotides apart. Just as ZFN and TALEN, this technology can be used to cause deletions as well insertions, relying on Homologous Recombination. But CRISPR/Cas9 utilises guide RNAs for target site recognition which makes this technology cheaper and easier to customize, while also being more effective (Doudna and Charpentier 2014; Jinek et al. 2012). Figure 1.2 shows the functional mechanism.

The most probable application would utilise a CRISPR/Cas9-mediated gene drive system inheriting a cargo gene to the vast majority of its offspring, which would burden the population's fitness. Although a CRISPR/Cas9-mediated gene drive could just as well be designed as a self-limiting drive, its capabilities would not fully be exploited if it is not applied as a self-sustaining drive.

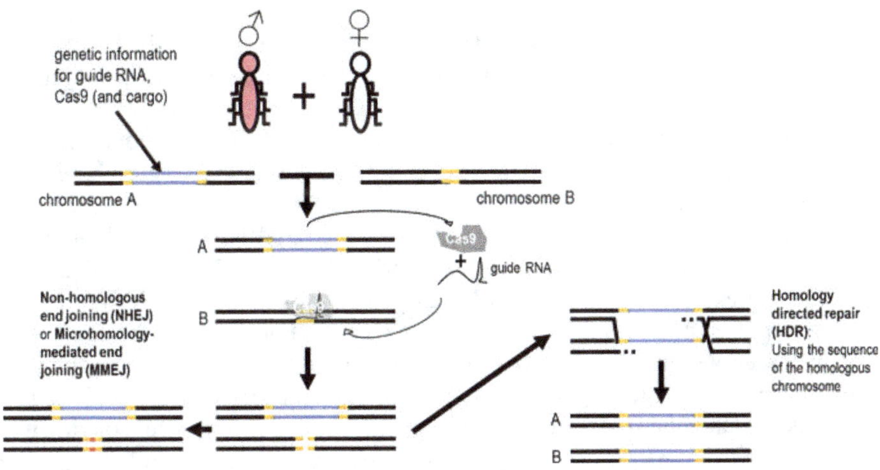

Fig. 1.2 Mechanism of CRISPR/Cas9-based gene drives. A gene drive organism carries the gene drive cassette (chromosome A) and mates with a wild type (chromosome B). The gene drive cassette expresses the CRISPR/Cas9 complex, which then cuts its recognition site defined by the gRNA on chromosome B. This cut then can either be repaired by Non-Homologous End Joining (NHEJ) or Microhomology-Mediated End Joining (MMEJ) creating a homing resistant allele, or by homology directed repair (HDR) copying the gene drive cassette into the cut region

Depth of CRISPR/Cas9-Gene Drive Intervention

Power and range of a CRISPR/Cas9-based gene drive system in comparison to other SPAGE techniques would score as high due to its overwhelming inheritance. But power and range of a drive system are additionally determined by the functionality of the cargo gene and its burden to the fitness of the population. Since the drive is self-sustaining for multiple generations its range is probably to be scored as high as well, causing the overall depth of intervention rating to be concomitantly high.

Intensity of CRISPR/Cas9-Gene Drive Intervention

Due to its non-Mendelian inheritance, this gene drive would be frequency independent. Therefore, this technology's necessary frequency and thus also the intensity of intervention scores as low in comparison to other SPAGE techniques.

Reliability of the CRISPR/Cas9-Gene Drive Technology

The probability of failure of the CRISPR/Cas9-gene drive technology compared to other SPAGE techniques, considering the current state of development is high. Key points of error encompass:

- the selection against the fitness burden which the cargo gene could pose
- Non-homologous end joining (NHEJ) and microhomology-mediated end joining (MMEJ)
  In contrast to homology-directed repair (HDR), NHEJ and MMEJ reduce the conversion rate and may cause resistance due to mutations, deletions etc. Depending on the genomic location, HDR vs. NHEJ efficiency could be as low as ~10% (Lin and Potter 2016). Usually whenever a cut is repaired by NHEJ the result is a drive resistant allele. To reduce these events CRISPR/Cas9 could be used to enhance HDR gene expression and repress NHEJ genes. This could be achieved by the inclusion of HDR genes and NHEJ repressor genes. Furthermore, the generation of nucleases creating sticky-end overhangs as opposed to blunt ends may optimize the repair in the target organism. The rate of HDR depends on the species, cell type, developmental stage, and cell cycle phase. For example faithful copying was achieved with up to 97% efficiency in mosquitoes but only 2% in fruit flies (Esvelt et al. 2014).
- Incomplete or imperfect copying during HDR
  If the deletion preserves the reading frame it leads to a homing-resistant allele (Marshall et al. 2017).
- Emergence of homing resistant alleles due to random target site mutagenesis.
  This is circumvented by engineering multiple attack loci for the CRISPR/Cas9-system in the genome reducing the chance of mutation in all attacked alleles.

However, very large populations—such as those of some insects—might require unfeasibly large numbers of gRNAs to prevent resistance (Ping 2017).

- Off-target-effects
  Unspecific binding of gRNA causes unintended insertions at different loci.
- On-target mis-insertions
  Unwanted genes or gene fragments are inserted into the target locus, instead or additionally to the desired genes.
- Sequence polymorphisms
  Resistance due to genetic variations within a species. To overcome this problem multiple gRNA variants can be added to the CRISPR/Cas cassette.
- Intragenomic interactions
  The distance of gRNA target sites may affect homing rates (Marshall et al. 2017).
- Maternal Effects
  Dominant maternal effects due to Cas9 deposits may cause resistance. Propagation of resistant individuals may be prevented by targeting essential genes (Noble et al. 2016). Homing and integration seems to occur in the germline. Upon fertilisation, if sufficient Cas9 (and gRNA) is in the cytoplasm of the female embryo which is homo- or heterozygous for the gene drive, maternal effects can occur. The CRISPR/Cas complex finds and cuts its recognition sites in the sperm's genome before the homologous female genome is close enough to be recruited for homologous recombination. Without a homologous template the cuts are then repaired by non-homologous end joining and thus arises a resistant allele. In such an event the number of gRNA variants is meaningless.
- errors in the release:

– phenotypic wild types carrying the non-functioning construct (would reduce the suppressive effect and could constitute a persistently gene drive-resistant subpopulation).

Vulnerability of the Target System Towards the CRISPR/Cas9-Gene Drive Technology

- adverse ecological effects due to the eradication of the population (depending on the fitness burden and ratio of released to native organisms)
- effects due to the spread to other populations (which becomes more probable the lower the fitness burden and the longer the gene drive is sustained in the ecosystem)

Possibilities for Limitation of Damage Caused by CRISPR/Cas9-Gene Drive Technology

It is not yet possible to make reliable statements about the effectiveness of options for limiting or reversing the changes caused by released CRISPR/Cas-based gene drives.

Different measures for the inhibition of their spread as well as for the inactivation of the functionalities induced by gene drives have already been proposed and also a first experiment in yeast has been undertaken (DiCarlo et al. 2015; Esvelt et al. 2014). However, a proof of their efficacy when used in insects or other higher organisms has not yet been established. In addition, there is still no possibility for the complete restoration of the natural gene sequence after the spread of RNA-guided gene drives.

Certain concepts to restrict the uncontrolled spread of a CRISPR/Cas9-mediated gene drive are:

- A reversal drive which could be used to overwrite a first drive, although it would have to be recoded to be immune to this first drives cutting. A third drive could then restore the original wild type sequence, although the cas9-gene and gRNAs would remain (Esvelt et al. 2014).
- Immunising drives could be used (pre-emptively) to render populations immune to another drive by recoding the sequences targeted by that drive (Esvelt et al. 2014).
- In a split drive the genomic locations for the components of the drive are separated in such a way that only a certain part of the information for a functionally active drive is inherited. This serves the local confinement of a gene drive (DiCarlo et al. 2015).
- A daisy chain drive is defined by a linear chain of interdependent drive elements on different genomic loci in which the first drive element is responsible for the duplication of the second, the second for the third etc. but the first drive element is not duplicated and therefore the whole drive systems successively gets lost over time (Noble et al. 2016).
- An overwriting drive (for restoring edited traits) was tested in yeast (DiCarlo et al. 2015) but not in higher organisms.

Summary of the Technology Characterisation

In Table 1.1, the here discussed SPAGE technologies are compared in certain characteristics as far as estimates are possible, considering information available in the literature. In the subsequent Table 1.2 the different SPAGE techniques and their evaluation resulting from the above represented technology characterisation are put together for better comparison.

As a prerequisite for further orientation on the impact and potential exposure of SPAGE, common features of these technological approaches were selected for a comparative technology characterisation (power and range) as well as an analysis of factors (traits) which influence its impact, spread and invasiveness.

As explained before, all SPAGE-Technologies are determined as of very high depth of intervention. In Table 1.2, the focus lies on a differentiation within the field of SPAGE technologies. Their general high depth of intervention is presupposed.

Table 1.1 SPAGE technique comparison

	Bi-sex RIDL	fsRIDL	X-Shredder	Killer-Rescue	Medea	UD	CRISPR/Cas
Resistance formation	Possible	Possible	Very unlikely	Possible	Possible	Unlikely	Likely
fitness	~56%	~56%	Unknown	Unknown	82.6%	Probably lower than medea	Probably low
Invasiveness	Very low	Very low	Moderate	High	Moderate	Low	High
Toxicity	Likely	Likely	None	Possible	None	None	None
Corrigibility	WT release	WT release	None	WT release	2nd generation drive	WT release	Reversal drive
Class	Suppression	Suppression	Suppression	Replacement	Replacement	Replacement	Replacement
Mode of Action	Toxin	Sex ratio/toxin	Sex ratio/chromo-somal disruption	Toxin/antidote	Toxin/antidote	Toxin/antidote	Heterozygote to Homozygote
Linkage loss	Unknown	Unknown	Unknown	Unknown	Unlikely	Unlikely	Likely
Estimated interspecies gene flow	Very likely	Very likely	More likely	Likely	Likely	Likely	Depends

Table 1.2 Comparative Technology Characterisation of SPAGE Techniques

SPAGE-Type	Intensity of the intervention	Depth of intervention technological power/range	Probability of failure	Discussed possibilities for corrective action/limitation of damage in case of failure
Killer rescue	High	Low/moderate	Low	Wild type release
Medea	High	Moderate/high	Low	2nd generation Medea drive
Two-locus underdominance	High	Moderate/low	Moderate	Wild type release
Bi-sex RIDL	High	High/low	Moderate	Wild type release
fsRIDL	High	High/moderate	Moderate	Wild type release
X-Shredder	High	High/high	Moderate	None
CRISPR/Cas9 Gene Drive	Low	High/high	High	Rescue drive

Such an internal differentiation is an important knowledge base for a differentiated risk management.

The comparative technology characterisation revealed that concerning reliability there are no remarkable differences, but there are differences especially in the spectrum of power and range which presumably lead to different levels of potential hazards and exposure. For instance, SPAGEs may employ different mechanisms to ensure their mode of inheritance. From simple lethality by toxic gene products (RIDL), through more or less intricate toxin-antidote systems as Medea, Underdominance, Killer Rescue to the biased segregation of sex chromosomes during meiosis (X-Shredder). An extreme potential with regard to power and especially range could be identified for endonuclease-based gene drives using the CRISPR/Cas9-system. Moreover, as for some other SPAGEs, its probability of failure is comparably high. The outstanding potential of CRISPR/Cas9-based gene drives was also illustrated by the assessment of the range based on invasiveness of different SPAGE-techniques according to their inheritance schemes. Along with power and range uncertainties and ignorance rise with (a) the extent of known unknowns regarding potential effects of known dependencies and relationships of the target species and possibly affected non-target species and (b) not yet determinable effects (unknown unknowns) due to unknown relationships or the inherent instability of genetic information which becomes more relevant with increasing numbers of gene drive-modified organisms.

In the light of the absence of proven options to (a) correct potential damage or (b) just to limit the inherently self-propagating mechanism of SPAGE, these properties reveal important 'reasons for concern' with regard to the requirements of the precautionary principle.

References

Ahrens, A., Braun, A., von Gleich, A., Heitmann, K., & Lißner, L. (2005). *Hazardous chemicals in products and processes—Substitution as an innovative process. Sustainability and innovation.* Heidelberg, Germany; New York, USA: Physica Verlag.

Akbari, O. S., Matzen, K. D., Marshall, J. M., Huang, H., Ward, C. M., & Hay, B. A. (2013). A synthetic gene drive system for local, reversible modification and suppression of insect populations. *Current Biology, 23*, 671–677. https://doi.org/10.1016/j.cub.2013.02.059.

Akbari, O. S., Chen, C.-H., Marshall, J. M., Huang, H., Antoshechkin, I., & Hay, B. A. (2014). Novel synthetic Medea selfish genetic elements drive population replacement in drosophila, and a theoretical exploration of Medea-dependent population suppression. *ACS Synthetic Biology, 3*, 015–928.

Akbari, O. S., Bellen, H. J., Bier, E., Bullock, S. L., Burt, A., Church, G. M., et al. (2015). Safeguarding gene drive experiments in the laboratory. *Science, 349*, 927–929. https://doi.org/10.1126/science.aac7932.

Alphey, L. (2014). Genetic control of mosquitoes. *Annual Review of Entomology, 59*, 205–224. https://doi.org/10.1146/annurev-ento-011613-162002.

Alphey, L., McKemey, A., Nimmo, D., Neira Oviedo, M., Lacroix, R., Matzen, K., et al. (2013). Genetic control of Aedes mosquitoes. *Pathogens and Global Health, 107*, 170–179. https://doi.org/10.1179/2047773213Y.0000000095.

Australian Academy of Science. (2017). Synthetic gene drives in Australia: implications of emerging technologies. https://www.science.org.au/files/userfiles/support/documents/gene-drives-discussion-paper-june2017.pdf. Accessed June 7, 2019.

Buchman, A., Marshall, J. M., Ostrovski, D., Yang, T., & Akbari, O. S. (2018). Synthetically engineered Medea gene drive system in the worldwide crop pest Drosophila suzukii. *PNAS, 115*, 4725–4730.

Burt, A. (2003). Site-specific selfish genes as tools for the control and genetic engineering of natural populations. *Proceedings of the Royal Society B: Biological Sciences, 270*, 921–928. https://doi.org/10.1098/rspb.2002.2319.

Burt, A., & Trivers, R. (2006). *Genes in conflict: The biology of selfish genetic elements*. Cambridge MA: Belknap Press/Harvard University Press.

Carareto, C. M. A., Kim, W., Wojciechowski, M. F., O'Grady, P., Prokchorova, A. V., Silva, J. C., et al. (1997). Testing transposable elements as genetic drive mechanisms using Drosophila P element constructs as a model system. *Genetica, 101*, 13–33.

Champer, J., Reeves, R., Oh, S. Y., Liu, C., Liu, J., Clark, A. G., et al. (2017). Novel CRISPR/Cas9 gene drive constructs in Drosophila reveal insights into mechanisms of resistance allele formation and drive efficiency in genetically diverse populations. bioRxiv. https://doi.org/10.1101/112011.

Commission of the European Communities. (2000). Communication from the commission on the precautionary principle. Brussels, Belgium.

Courtier-Orgogozo, V., Morizot, B., Boëte, C. (2017). Agricultural pest control with CRISPR-based gene drive: time for public debate: Should we use gene drive for pest control? EMBO Report e201744205. https://doi.org/10.15252/embr.201744205.

Craig, G. B. J., Hickey, W. A., & VandeHey, R. C. (1960). An inherited male-producing factor in *Aedes aegypti*. *Science, 132*, 1887–1889. https://doi.org/10.1126/science.132.3443.1887.

Curtis, C. F. (1968). Possible use of translocations to fix desirable genes in insect pest populations. *Nature, 218*, 368–369.

Deredec, A., Burt, A., & Godfray, H. C. J. (2008). The population genetics of using homing endonuclease genes in vector and pest management. *Genetics, 179*, 2013–2026. https://doi.org/10.1534/genetics.108.089037.

DiCarlo, J. E., Chavez, A., Dietz, S. L., Esvelt, K. M., & Church, G. M. (2015). Safeguarding CRISPR-Cas9 gene drives in yeast. *Nature Biotechnology, 33*, 1250–1255. https://doi.org/10.1038/nbt.3412.

Doudna, J. A., & Charpentier, E. (2014). The new frontier of genome engineering with CRISPR-Cas9. *Science, 346*. https://doi.org/10.1126/science.1258096.

EC DIRECTIVE 2001/18/EC of the European Parliament and of the Council of 12 March 2001 on the deliberate release into the environment of genetically modified organisms and repealing Council Directive 90/220/EEC. https://www.epa.ie/pubs/legislation/geneticallymodifiedorganismsgmo/2001-18%20Directive_consolidated.pdf.

Emerson, C., James, S., Littler, K., & Filippo, R. (2017). Principles for gene drive research. *Science, 358*, 1135–1136. https://doi.org/10.1126/science.aap9026.

Esvelt, K. M., Smidler, A. L., Catteruccia, F., Church, G. M. (2014). Concerning RNA-guided gene drives for the alteration of wild populations. *eLife, 3*, e03401. https://doi.org/10.7554/eLife.03401.

European Commission, Scientific Advice Mechanism. (2017). New Techniques in agricultural biotechnology. https://ec.europa.eu/research/sam/index.cfm?pg=agribiotechnology

European Environment Agency. (2002). Late lessons from early warnings: The precautionary principle 1896–2000. *Environmental Issue Report, 22*, 1–211.

European Parliament Think Tank. (2015). The precautionary principle: Definitions, applications and governance. http://www.europarl.europa.eu/RegData/etudes/IDAN/2015/573876/EPRS_IDA(2015)573876_EN.pdf. Accessed June 7, 2019.

Fischer, E., Jones, J., & von Schomberg, R. (2006). *Implementing the precautionary principle—Perspectives and prospects*. Bodmin, Cornwall, Great Britain: MPG Books Ltd.

Galizi, R., Doyle, L. A., Menichelli, M., Bernardini, F., Deredec, A., Burt, A., et al. (2014). A synthetic sex ratio distortion system for the control of the human Malaria mosquito. *Nature Communications, 5*, 3977. https://doi.org/1038/ncomms4977.

Gantz, V. M., & Bier, E. (2015). Genome editing. The mutagenic chain reaction: A method for converting heterozygous to homozygous mutations. *Science, 348*, 442–444. https://doi.org/10.1126/science.aaa5945.

Gantz, V. M., Jasinskiene, N., Tatarenkova, O., Fazekas, A., Macias, V. M., Bier, E., et al. (2015). Highly efficient Cas9-mediated gene drive for population modification of the malaria vector mosquito Anopheles stephensi. *Proceedings of the National Academy of Sciences, 112*, E6736–E6743. https://doi.org/10.1073/pnas.1521077112.

Gene Watch UK. (2013). Genetically Modified (GM) Olive flies: A credible pest management approach? https://www.genewatch.org/uploads/f03c6d66a9b354535738483c1c3d49e4/GMolivefly_GWbriefing_fin2.pdf.

Giese, B., & von Gleich, A. (2015). Hazards, risks, and low hazard development paths of synthetic biology. In B. Giese, C. Pade, H. Wigger, & A. von Gleich (Eds.), Synthetic biology—Character and impact. Heidelberg: Springer International Publishing.

Gokhale, C. S., Reeves, R. G., & Reed, F. A. (2014). Dynamics of a combined medea-underdominant population transformation system. *BMC Evolutionary Biology, 14*, 1–9. https://doi.org/10.1186/1471-2148-14-98.

Gould, F., Huang, Y., Legros, M., & Lloyd, A. L. (2008). A Killer-Rescue system for self-limiting gene drive of anti-pathogen constructs. *Proceedings of the Royal Society B: Biological Sciences, 275*, 2823–2829. https://doi.org/10.1098/rspb.2008.0846.

Hastings, I. M. (1994). Selfish DNA as a method of pest control. *Philosophical Transactions of the Royal Society of London. Series B, 344*, 313–324.

Helleu, Q., Gérard, P. R., & Montchamp-Moreu, C. (2015). Sex chromosome drive. *Cold Spring Harbor Perspectives in Biology, 7*, a017616. https://doi.org/10.1101/cshperspect.a017616.

Hochkirch, A., Beninde, J., Fischer, M., Krahner, A., Lindemann, C., Matenaar, D., et al. (2017). License to kill? Disease eradication programs may not be in line with the convention on biological diversity. *Conservation Letters, 11*, 1–6. https://doi.org/10.1111/conl.12370.

Jinek, M., Chylinski, K., Fonfara, I., Hauer, M., Doudna, J. A., & Charpentier, E. (2012). A programmable dual-RNA-guided DNA endonuclease in adaptive bacterial immunity. *Science, 337*, 816–821. https://doi.org/10.1126/science.1225829.

Larracuente, A. M., & Presgraves, D. C. (2012). The selfish segregation distorter gene complex of *Drosophila melanogaster*. *Genetics, 192*, 33–53. https://doi.org/10.1534/genetics.112.141390.

Ledford, H. (2015). CRISPR, the disruptor. *Nature, 522*, 20–24. https://doi.org/10.1038/522020a.

Lin, C. C., & Potter, C. J., 2016. Non-Mendelian dominant maternal effects caused by CRISPR/Cas9 transgenic components in *Drosophila melanogaster*. *G3: Genes, Genomes, Genetics, 6*, 3685–3691. https://doi.org/10.1534/g3.116.034884.

Lindholm, A. K., Dyer, K. A., Firman, R. C., Fishman, L., Forstmeier, W., Holman, L., et al. (2016). The ecology and evolutionary dynamics of meiotic drive. *Trends in Ecology & Evolution, 31*, 315–326. https://doi.org/10.1016/j.tree.2016.02.001.

Linkov, I., Trump, B. D., Anklam, E., Berube, D., Boisseasu, P., Cummings, C., et al. (2018). Comparative, collaborative, and integrative risk governance for emerging technologies. *Environment Systems and Decisions, 38*, 170–176.

Marshall, J. M., & Akbari, O. S. (2015). Gene drive strategies for population replacement. *Chapter 9 in Genetic control of Malaria and Dengue*. https://doi.org/10.1016/B978-0-12-800246-9.00009-0.

Marshall, J. M., Buchman, A., Sánchez C., Akbari, O. S. (2017). Overcoming evolved resistance to population-suppressing homing-based gene drives. *Nature Scientific Reports, 7*, 1–12. https://doi.org/10.1038/s41598-017-02744-7.

National Academies of Sciences. (2016). Gene drives on the horizon: Advancing science, navigating uncertainty, and aligning research with public values. Washington, DC: The National Academies Press. https://doi.org/10.17226/23405.

Newton, M. E., Wood, R. J., & Southern, D. I. (1976). A Cytogenetic analysis of meiotic drive in the mosquito, *Aedes Aegypti* (L.). *Genetica, 46*, 297–318.

Noble, C., Adlam, B., Church, G. M., Esvelt, K. M., Nowak, M. A. (2018). Current CRISPR gene drive systems are likely to be highly invasive in wild populations. *eLife, 7*, e33423. https://doi.org/10.7554/eLife.33423.

Noble, C., Min, J., Olejarz, J., Buchthal, J., Chavez, A., Smidler, A. L., et al. (2016). *Daisy-chain gene drives for the alteration of local populations.*. https://doi.org/10.1101/057307.

Owen, R., Crane, M., Grieger, K., Handy, R., Linkov, I., & Depledge, M. (2009). *Strategic approaches for the management of environmental risk uncertainties posed by nanomaterials—Nanomaterials: Risks and benefits*. Dordrecht, Netherlands: Springer.

Ping, G. (2017). Invasive species management on military lands: Clustered regularly interspaced short palindromic repeat/CRISPR-associated protein 9 (CRISPR/Cas9)-based gene drives. U.S. Army Engineer Research and Development Center (ERDC) Environmental Laboratory.

Reeves, R. G., Bryk, J., Altrock, P. M., Denton, J. A., & Reed, F. A. (2014). First steps towards underdominant genetic transformation of insect populations. *PLoS ONE, 9*, e97557. https://doi.org/10.1371/journal.pone.0097557.

Regalado, A. (2017). Farmers Seek To Deploy Powerful Gene Drive. https://www.technologyreview.com/s/609619/farmers-seek-to-deploy-powerful-gene-drive/URL. Accessed December 13, 2017.

Serebrovskii, A. S. (1940). On the possibility of a new method for the control of insect pests. *Zool Zhurnal, 19*, 618–630.

Silver, L. M. (1993). The peculiar journey of a selfish chromosome: mouse t-haplotypes and meiotic drive. *Trends in Genetics, 9*, 250–254. https://doi.org/10.1016/0168-9525(93)90090-5.

Sinkins, S. P., & Gould, F. (2006). Gene drive systems for insect disease vectors. *Nature Reviews Genetics, 7*, 427–435. https://doi.org/10.1038/nrg1870.

Steinfeldt, M., von Gleich, A., Petschow, U., & Haum, R. (2007). *Nanotechnologies Hazards and Resource Efficiency*. Heidelberg, Germany: Springer.

Sweeny, T. L., & Barr, A. R. (1978). Sex ratio distortion caused by meiotic drive in a mosquito, *Culex pipiens* L. *Genetics, 88*, 427–446.

Taylor, D. R. (1994). The genetic basis of sex ratio in *Silene alba* (= *S. latifolia*). *Genetics, 136*, 641–651.

The Rio Declaration on Environment and Development. (1992). https://www.unesco.org/education/pdf/RIO_E.PDF (accessed 7 June 2019).

United Nations. (2000). Cartagena protocol on biosafety to the convention on biological diversity. Secretariat of the Convention on Biological Diversity, Montréal, Canada, https://www.cbd.int/doc/legal/cartagena-protocol-en.pdf.

Webber, B. L., Raghu, S., & Edwards, O. R. (2015). Opinion: Is CRISPR-based gene drive a biocontrol silver bullet or global conservation threat? *Proceedings of the National Academy of Sciences U S A, 112*, 10565–10567. https://doi.org/10.1073/pnas.1514258112.

2

Tipping Points and Tipping Elements

Broder Breckling and Arnim von Gleich

The Relevance of Tipping Points to Understand Implications of Deliberate Release of Self Propagating Artificial Genetic Elements (SPAGE)

Tipping points and tipping elements, phase transitions and similar critical phenomena are widely discussed in scientific as well as socio-economic contexts as components to understand unforeseen far reaching changes and critical transitions from one stage into another in complex systems caused by small perturbations or gradual changes. For the risk assessment of self-propagating artificial genetic elements in self-sustaining wild populations of animals or plants, it is crucial to understand, where tipping elements could become relevant, how they could be anticipated and to what extent surprises and unexpected effects might occur.

Tipping points are an issue studied in a wide scope of scientific disciplines. According to Lenton (2013) the term tip point "was first introduced in sociology in the 1950s to describe the percentage of nonwhite residents in a US city neighbourhood that would trigger a white flight" (p. 2 with reference to Grodzins 1957). There exists a long scientific tradition of research regarding tipping points in physics (Domb and Green 1972 ff) and a tradition in ecosystems theory and application as well: The concept of tipping points helped to understand severe transitions in system structures (Scheffer et al. 2001; de Yong et al. 2008). Actually tipping points are intensively discussed as a component of 'systemic risks' regarding climate change (Lenton et al. 2008; Lenton 2011a, b, 2012) and turmoil in financial markets (Sornette 2003). By signing the Paris Agreement (United Nations Climate Change 2018)

B. Breckling (✉)
Landscape Ecology, University of Vechta, Vechta, Germany
e-mail: broder.breckling@uni-vechta.de

A. von Gleich
Department of Technological Design and Development, Faculty Production Engineering, University of Bremen, Bremen, Germany

an attempt to avoid catastrophic disruptions by touching tipping points in the climate system[1] even found its way into international regulation. It was claimed that limiting the raise of the global average temperature to 2 °C (or better to 1.5 °C) would keep climate change within the normal range of climate variation (cf. Nordhaus 1975; Rijsberman and Swart 1990; Randalls 2010; Lenton 2011b[2]).

Tipping points may be rare,[3] but when they are touched they often lead to catastrophic consequences.[4] Lack of knowledge, surprises, rare extreme events with severe consequences (called 'black swans' Taleb 2007) combine the debate about tipping points with the debate about precaution and the precautionary principle.

Tipping points refer to situations, where exceeding a particular (known—or previously unknown) threshold relates to subsequent and frequently self-amplifying changes over time, which lead to different systems states or even overall changes in the organization of a system. For biological systems, Schwegler (1981) developed a typology (see Fig. 2.1). There are numerous tipping points for which we already know the mechanism (the switch, the tipping element) and the triggered dynamics (bifurcations, phase transitions etc.). Regarding the triggered dynamics there is a wide scale of important differences in magnitudes. When positive feed-back processes are triggered, there might be changes that are difficult or even impossible to reverse, that do not allow recovery. Taleb et al. (2014) call this phenomenon "the risk of ruin" (p. 2f.).

Tipping points refer to critical systems stages, where smallest impulses or gradient shifts can trigger far-reaching consequences. A tipping point delimits domains of different dynamic behaviour of a system. When dealing with tipping points, 1. initial states/dynamics, 2. impulses, 3. tipping elements (structures or functions that can switch in the system, control parameters), 4. tipping mechanisms (swings, transitions, bifurcations, phase changes, etc.) and 5. the new states or dynamics are to be considered. Many of the aspects relevant to anticipate dynamic processes may be incompletely known or even unknown. Though tipping phenomena are part in everyday life and we are familiar with many of them, there are also surprising tipping phenomena with more or less far-reaching consequences. For those with the spatially and temporally highest impact, scientific approaches are required for adequate handling

[1]For instance, a shutdown of the large ocean circulation or a massive permafrost melting.

[2]Lenton however argued that "Yet, no actual large-scale threshold (or 'tipping point') in the climate system (of which there are probably several) has been clearly linked to 2 °C global warming" and he adds additional perturbations as possible triggers of tipping points: "distributions of reflective (sulfate) aerosols, absorbing (black carbon) aerosols, and land use could be more dangerous than changes in globally well-mixed greenhouse gases" (Lenton 2011b, p. 451).

[3]"The traditional deterministic chaos, for which the butterfly effect was named, applies specifically to low dimensional systems with a few variables in a particular regime. High dimensional systems, like the earth, have large numbers of fine scale variables. Thus, it is apparent that not all butterfly wing flaps can cause hurricanes. It is not clear that any one of them can, and, if small perturbations can influence large scale events, it happens only under specific conditions where amplification occurs" (Taleb et al. 2014, p. 12).

[4]"Passing a tipping point … is typically viewed as a 'high-impact low-probability' event" Lenton (2011b, p. 201).

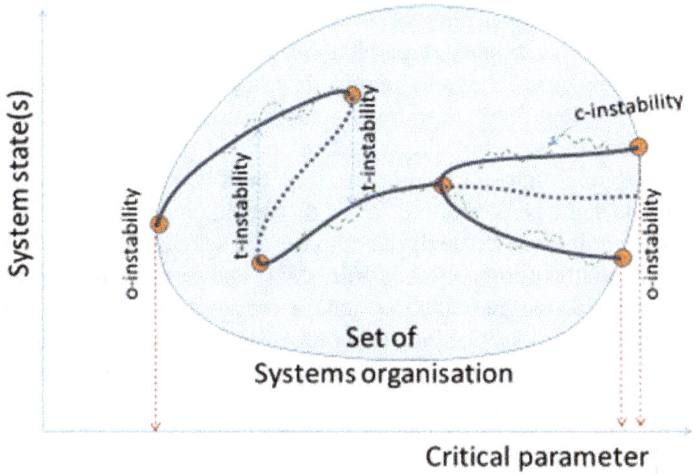

Fig. 2.1 Schwegler (1981) and (1985) introduced a formal stability concept which distinguishes transitions that are to be seen as tipping points (re-drawn and adapted). *c-instability* (named after the climax state of ecosystems) is the domain of system states that can be reached through parameter variations without discontinuities in the system development. This includes the passing of bifurcation points. A *t-instability* occurs, if a transition between segregated branches of stable state occurs as it is the case in hysteresis. *O-instability* means a transition, which is irreversible and leads to a loss of systems organisation

strategies. Along with advances in the theory of complex dynamic systems (homeostasis, self-organisation, emergence, chaos theory, synergetics, operations research), research has increasingly focused on problems of unpredictability, or the attempt to prepare for possible surprises through non-linear behaviour at tipping points. Areas of such precautionary research on tipping points are e.g. medicine, psychology, psychiatry, ecosystem theory, meteorology and research on climate change as well as social sciences and economics, especially business and finance. In this context, a separate research direction has emerged that deals with the precautionary question of the extent to which a threatening approach to a tipping point with serious consequences can be detected in good time for possible countermeasures (early warnings). Our work on possible tipping points in connection with research and development, regulation and release of gene drives can be classified here. We are primarily concerned with tipping points in technology development (technology characterisation), in the (agricultural) ecosystems that can be expected to be affected (vulnerability analysis) and in the areas of technology governance and acceptance/acceptability. The focus is on 'reasons for high concern' that can justify precautionary measures.

Perturbations touching tipping points may trigger completely unexpected behaviour of the system, confronting us with surprises. Such surprises are known from dealing with (socio-) ecological (Holling 1978; Filbee-Dexter et al. 2017), socio-economical (Kim and Mauborgne 2003; Helbing 2012) and socio-technical systems (Taleb 2007). Thus the interest of research on tipping points not only lies in the identification and description of different characteristics of tipping points but more and more in precautionary research about the possibilities and limitations to timely identify indications, that a system is approaching a tipping point. There is a growing literature about 'early-warning signals for critical transitions' in complex dynamical systems (Scheffer et al. 2009; Lenton 2011a). Its focus lies on system dynamics and tipping elements. Important aspects of tipping elements are thresholds, bifurcations and basins of attraction. If these elements and their consequences are already known and can be analysed empirically, scientific approaches are possible based on statistics (probabilistic forecasting) and on dynamic modelling (see Table 1 in Lenton 2011a). Identified as generic 'early warning signals' are certain courses of system behaviour like increasing time for recovering from perturbations, increasing autocorrelation, increasing variance, increasing skewness and flickering (ibid., see also de Yong et al. 2008; Veraart et al. 2012; Andersen et al. 2009; Carpenter et al. 2011 and regarding financial markets Wanfeng et al. 2010 and Gatfaoui and de Peretti 2019).

Still more challenging is the identification of tipping elements of which the consequences are still unknown because they did not yet happen or should not happen at all. A possible approach is the analysis of the dynamics and the structure (architecture) of the system. Regarding system dynamics, the above mentioned early warning signals of system behaviour may be helpful to identify even not yet known bifurcations or basins of attraction. It is an aim of the modelling efforts as described in Chap. 5 to contribute to such an approach. Regarding system architecture, a structural vulnerability analysis taking into account results of research on 'resilient systems' should be helpful (see Chap. 3: "Analysis of Vulnerability of Ecological Systems"). Taleb et al. for instance identified 'the connectivity of a system' as a potential for the 'propagation of harm' and they add that our socio-technological and socio-economic global systems are extremely interconnected especially by transportation and information technology ('the global connectivity of civilization', Taleb et al. (2014, p. 4)). For them 'invasive species' and 'rapid global transmission of diseases' are among the severest historical consequences. They recommend "boundaries, barriers and separations that inhibit propagation of shocks" (p. 10). The character of interconnections within systems is indeed an important aspect of vulnerability or resilience, others are structural and functional diversity and redundancy, buffers and stocks that increase resilience or the lack of which may trigger bifurcations or phase transitions (von Gleich and Giese 2019).

The capability of triggering bifurcations, phase transitions, alternative dynamic regimes or alternative stable states is not limited to the state and structure of the affected system. It is as well depending on the character and magnitude of the perturbations. And regarding gene drives we are dealing with self-amplifying perturbations. We focus on ambitions to develop organisms with a modification that facilitates gene drives. A gene drive is a genomic construct which enables a higher rate of inheritance of the construct than those occurring naturally as a result of Mendelian inheritance (Burt 2003). The character and magnitude of perturbation caused by gene drives is dependent on the one side on the implemented functionality, which can be very powerful and on the other hand on the extremely extended exposure (see Chap. 1

Technology Characterisation). Self-propagating exposure magnifies the opportunities to touch otherwise unreachable tipping points and thus increases the probability of the occurrence of ‘black swans’, of extremely improbable, “unforeseen and unforeseeable events of extreme consequences” (Taleb et al. 2014, p. 1). This is the reason why Taleb et al. (2014) request “more precaution about newly implemented techniques, or larger size of exposures” (Taleb et al. 2014 p. 7). Regarding tipping points, the expectable impact of technological interventions into socio-ecological systems can be identified by the method of a structural vulnerability analysis, with its focus on weak points independent of certain perturbations (see Chap. 3: “Analysis of Vulnerability of Ecological Systems”). An important element should be the test of model processes. Whenever an apparent trend to a significant alteration in a given context occurs, it is reasonable to ask in which way tipping point dynamics could be involved in order to understand ongoing and coming developments.

We already mentioned that tipping points occurred and may occur in socio-technological systems. With regard to molecular biotechnological developments we are apparently in an accelerated innovation phase, where rapid transitions occur, where new and previously inaccessible technical options are on the way to become feasible. The construction and especially the release of Self-Propagating Artificial Genetic Elements (SPAGE) in wild living populations of organisms is in itself touching a tipping point because of the new quality of technical power and the intended self-propagating exposure.

To outline how the understanding of the involvement of tipping points contributes to a rational understanding of the ongoing developments, we first discuss the significance of tipping point dynamics in exemplary contexts well established in the scientific discourse. Therefore, we summarize frequently used formal types of tipping points known from the analysis of dynamic systems. Then we take a look in which regard the concept of tipping points helps to understand the potential transitions that are involved in possible applications of SPAGE. For this purpose, we use a systems hierarchy approach to address the different organisation levels which could be influenced by tipping point transitions.

Conceptual Background: Phase Transitions in Dynamic Systems

We first take a look at examples of phase transitions in different scientific contexts. This underlines that tipping point dynamics are not a singular phenomenon in ecology but a fundamental topic in science dealing with dynamic systems, for which a considerable body of theory is available that helps to understand possible risks and supports the development of adequate handling strategies in various contexts. With regard to tipping points it became obvious that they are of specific importance in the management of ecological systems and that concept transfers between different disciplines helped in understanding case-specific details.

Examples in Different Scientific Domains

Phase transitions that take place when exceeding critical stages in various physical, ecological and even socio-economic contexts encompass scales from the global level down to small-scale systems. We summarize some of the most widely studied systems where tipping points are crucial for understanding.

Biosphere Ecology: Global Climate Change

As already mentioned tipping points are mainly discussed in the current scientific as well as public debate with regard to the global climate system and the changes that global transitions may bring for regional scales, land use, and security of coastal zones (Bindoff et al. 2014; Parmesan and Yohe 2003; Harley et al. 2006). It is assumed, that the global climate system represents a steady state with limited random shifts close to an equilibrium which depends on the overlay of different feedback mechanisms. Disturbances which might raise the global average temperature above about two degrees is currently assumed to activate processes that lead to further warming and potential unmanageable transitions in the global climate system (Lenton 2011b). These processes are referred to as tipping elements in the global climate system (Lenton et al. 2008). Considerable efforts in research as well as in political administration on various levels are on the way to identify and to quantify these potentially critical thresholds and avoid the exceedance of tipping points beyond which undesirable self-amplifying traits gain an extent and intensity that might go beyond available management capacities and lead to an overall state of the global climate system which provides less management options than the currently prevailing state of the global climate system.

There are several critical components considered, which may be involved in the tipping point dynamics within global climate system (Fig. 2.2): A particular extent of global warming could reduce polar and high mountain ice cover. This would significantly reduce the earth's albedo, i.e. the light reflection of the earth and the absorption of solar irradiance would increase. This could lead to further melting of ice leading to further warming until a higher equilibrium would be reached (Notz 2009; Lenton 2012). Another tipping element could be the thawing of permafrost soils, driving the system in the same direction. The onset of bacterial activity in water saturated and previously frozen soil with a high content of dead organic material would lead to the release of large amounts of CO_2 as well as methane (CH_4) to the atmosphere. Methane is about 34 times higher in efficiency as a greenhouse gas than CO_2 (Dean 2018; Biskaborn et al. 2019). Another tipping element could be the release of methane hydrates in the ocean shelf marginal areas, which might become unstable when water temperatures increase (Kvenvolden 1988; United Nations Climate Change 2014). Global climate research attempts quantifications of these and additional other tipping elements as well as counteracting processes to come up with

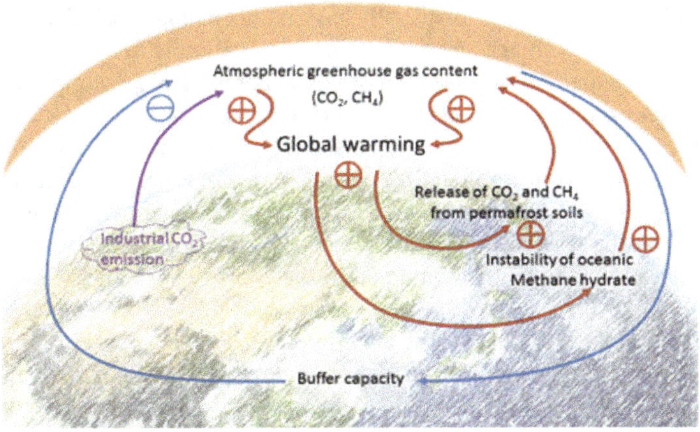

Fig. 2.2 The global climate is governed by a variety of interlinked feedback processes, among of which are the release of carbon dioxide (CO_2) and methane (CH_4) from thawing permafrost soils, and methane release from potentially instable methane hydrates under sediments on the ocean floor as a result of increasing temperatures. In the discussion on global warming, it is considered as probable, that tipping points exist, beyond which there would be the onset of a self-amplifying shift towards a different equilibrium of globally higher temperatures. Though a global trend is generally widely detected, there is uncertainty remaining how the different processes are interlinked and how a new equilibrium would stabilise

an integrative picture in which tipping point dynamics appear to play a crucial role for the understanding of long-term and large-scale dynamics.

Tipping point dynamics have a crucial role also in many smaller scale processes.

Fire Ecology

In many ecosystems, fire is an important structuring factor. In landscapes where accumulated organic matter is sufficiently abundant and dry, fire events of various orders of magnitude can take place as a regular part of the systems dynamics. Fire ignition can have natural reasons (e.g. lightning, electrostatic discharges) or emerge due to anthropogenic influence. The attempt to understand the rapid transition process from heat mediated, self-amplifying transition of organic material to ashes led to the development of a specific scientific discipline: fire ecology—with an elaborated body of empirical as well as theoretical findings (e.g. Goldammer 2012). This encompasses the insight, that large-scale natural fire are a regular component of those landscapes, in which a pronounced seasonal change between draught and humidity facilitates an accumulation of organic material. Fire events become the more probable to expand, the higher the amount of accumulated combustible material is. In these landscapes, fire leads to a characteristic vegetation pattern with larger even-aged plant stands, which result from the temporal synchrony of re-growth starting

(Johnson 1996). Young vegetation usually has a reduced inflammability which is successively increasing with age. An interesting implication is, that an anthropogenic suppression of the fire outbreak without reducing combustible material (e.g. through prescribed fires) might increase the risk and the extent of large fires and thus the subsequent damage potential of single events (Piñol et al. 2005). The critical point that triggers the phase transition in fire ecology is apparently the randomly occurring igniting event. How far the single event expands across larger parts of the landscape is influenced by the particular vegetation structure and the extent of available combustible material which was accumulated in the ecological system (Chandler et al. 1983).

Fire ecology can also be used as an example to explain that the tipping point concept is helpful on different levels of the same context: Landscape analysis can bring about insight in a further tipping element. In random pattern of inflammable and non-inflammable sites a density threshold can exist beyond which the spatial range of a fire tremendously expands (see the explanation below under percolation).

Population Ecology: Outbreaks and Epidemics

The population outbreak dynamics that can be observed in some biota, e.g. in the form of plankton blooms appears to be phenomenologically related. There are modelling approaches departing from similar formal descriptions. In aquatic systems there is the possibility of a sudden extreme quantitative increase of particular species with dramatic consequences for other organic ecosystem components. While usually being embedded in stabilizing nutrient capacity limitations or predator–prey interactions, an escape from these limitations can occur under certain conditions and give rise to extremely increasing, several orders of magnitude higher population densities (Hallegraeff 1993). In these systems, a precise tipping point is not easy to determine, though in dynamic model representations the behaviour can be studied (e.g. Huppert et al. 2002).

The outbreak of epidemics and biological invasions follows comparable patterns and is frequently analysed with similar formal approaches (Earn et al. 2000).

Aquatic Ecology: Alternative Stable States in Shallow Lakes and Other Ecological Systems

During the 1970s it was discovered, that in temperate climates shallow lakes can shift between two different types of organisms that dominate the species composition (biocenosis). Under otherwise comparable external conditions, the biocenosis can be either dominated by ground-rooted macrophytes or alternatively by plankton algae. After extensive limnological studies (Scheffer et al. 1993; Scheffer and van Nes 2007) the conclusion was drawn, that it is possible to force transitions between these different states. Such a transition can be induced by considerably increasing or decreasing the nutrient load, in particular phosphate: A strong eutrophication allows the plankton to thrive to an extent that the macrophytes are shaded out and vanish from the system. The other way round, a restoration of the macrophyte dominance requires an extensive removal of nutrients—far more than required to establish the algae-dominated stage. It occurs, when the nutrient limitation is sufficiently pronounced so that algae density becomes sufficiently thin so that shading of macrophyte re-growth is no more efficient.

This type of switching between alternative stable states is known also from a variety of other, in particular physical, systems (e.g. magnetism, Jiles and Atherton 1986). The involved direction-dependent transitions are called hysteresis (Mayergoyz 2012 see also Fig. 2.9). In an ecological context, the shallow lakes were one of the first examples in ecology being studied and successfully modelled on a quantitative basis. Meanwhile, this example encouraged additional research for comparable phenomena in other ecosystems. Today, quite a number of other ecological processes is known that follow an according pattern. Scheffer et al. (2001) provide a list of studies describing this type of ecological transitions.

Concepts and Applications in Dynamic Theory: Important Forms of Phase Transitions Where Tipping Points Mark Domain Boundaries

The previously described examples have been modelled in formal structures which can be assigned to specific forms of transitions. While the global climate system requires an overlay of a number of different counteracting processes of different dynamic forms, the conceptual understanding of other transition processes is less complex and it is easier to identify a limited number of dominating drivers. Knowing them informs about the repertoire of approaches that could play a role also with regard to SPAGE. The following dynamic paradigms are frequently employed to understand tipping point dynamics. A number of approaches was developed for mathematical modelling of physical and chemical as well as for ecological and social systems, which capture different tipping point situations. The approaches illustrate, that the metaphor of a tipping point summarises different forms, applicable to particular situations.

Balance and Seesaw

The most simplistic form of a tipping point model follows the idea of a balance or a seesaw (Fig. 2.3). Putting increasing amounts of weight on the lighter side leads to an equilibrium point on which both sides reach equivalence. Exceeding this point causes the system to flip over. A fundamental characteristic of tipping points is apparent in such a system: For a distinction to which side the seesaw will tilt, there are large domains where changes in load have a limited effect. This is different precisely at

Fig. 2.3 An infinitely small impact can be linked to an infinitely large effect. This is characteristic for a seesaw precisely at the tipping point .

the tipping point. At the tipping point, infinitesimal influences can be decisive, to which side the seesaw tilts. Prigogine and Stengers (1984) argue for a generalization of the effect: that at the tipping point, the principle of causality, i.e. the consistent connection of quantifiable cause and effect, does not hold, because for this point no finite relation of cause and effect can be specified. Precisely at the border between two different dynamic domains infinitesimally small influences can trigger effects of any order or magnitude—depending on the systems specification. With regard to SPAGE, it can be argued, that resulting parameter changes can influence ecological balances and shift systems states towards new equilibria. Models (see Chap. 5) can illustrate such an effect.

Domino Effects: Iteration

Spatially extended iterative processes are a significant domain for tipping point dynamic studies. The domino effect (Fig. 2.4) can be seen as a simplistic model of such an iterative phenomenon. In systems, where long cause-effect chains play a role, small-scale local events can affect their neighbourhoods in a way that the

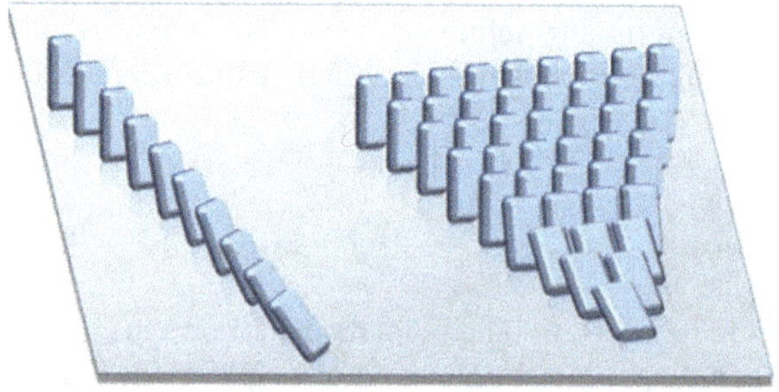

Fig. 2.4 Falling domino pieces are a frequently used metaphor and abstraction of interlinked cause-effect chains. In an ecological context, it has to be emphasised, that this kind of chain- or network-propagation is not necessarily linear but can in principle also follow amplification dynamics including exponential processes

change in a local state is transmitted to the surroundings. If the latter is capable to perform the same transition, large-scale macroscopic results can be caused by a local tip. A chain, where the fall of a single domino stone causes the next one to fall demonstrates a linearly proceeding dynamic. It is equally possible that an array is exposed in a way that one stone interacts with two or more others. Then specific exponential processes can emerge.

With some extensions, critical mass dynamics can be described by such an approach: The atoms of some radioactive elements decay at a random rate and emit particles and/or radiation. If they hit other kernels in the surrounding, they can induce additional fissions from which more particles are emitted. In a sub-critical mass, the spontaneous decay induces on average less that one additional fission in the surrounding. If, however, the density of unstable atoms within the energetic range of emitted particles is sufficiently high, an exponential increase of radioactive decays would emerge as a chain reaction and cause a nuclear explosion. The critical mass is specified by the probabilistic condition that on average, the fission of a single atom must cause more than one additional fission event (Holdren and Bunn 2002).

It is obvious to identify for such a case a tipping point with the critical mass. Other models of cascading effects follow comparable dynamics. This is in particular the case for outbreak and epidemic processes which occur in population systems. Its relevance for the SPAGE context results from its importance for inter-generational self-amplification.

Excitable Media

A variety of chemical reactions (e.g. Belousov–Zhabotinsky reaction, Briggs–Rauscher reaction) as well as physiological (e.g. neuron excitation, tachycardia and ventricular fibrillation) and ecological systems reaction (e.g. some lichen growth forms and vegetation pattern) can be described by the excitable media type of interaction. It is one of the basic forms of a pattern-generating mechanism in spatially structured self-organising systems. In these spatially structured systems, three different states can occur locally: 1—a locally *excitable* state which can persist for long times until it is activated by 2—an *excited* state occurring in direct neighbourhood. Then 1 also transits to the excited state. In analogy to a chain reaction thus the excitement can spatially propagate. The excited state persists for a particular time span and then turns into 3—a *refractory* state, which is no more excitable for a specific time span required for a regeneration of the excitable state. The local transitions are usually modelled on a grid to study the emergence of spatial patterns (Fig. 2.5). Though the underlying conception is rather simplistic, an according parametrization approach facilitates possibilities to describe a wide variety of pattern generating processes. A large body of literature exists, theoretical studies (Holden et al. 2013; Meron 1992) as well as applied studies (Kapral and Showalter 2012; Mosekilde and Mouritsen 2012). Fire ecology can be considered as a special case of an excitable media system, where the excited phase is comparatively very short, and the refractory phase by far longer. Also many processes of invasions or epidemics and a number of predator prey

Fig. 2.5 Excitable media pattern:
The spatially distributed alteration of a locally excitable, excited and refractory state (equally: susceptible, infected, recovering) is used to describe a wide variety of biological as well as social pattern generating processes. Among typical spatial pattern that can occur, are coherent spiral waves as well as transient random configurations. Here, a simulation on a hexagonal grid is shown where the transition from an excited to an excitable cell occurs at a probability of 0.5, while remaining for 3 iterations in an excited, and 4 iterations in a refractory state before returning to excitable. The grayscale indicates the age of the cells after the most recent excitation with white as the most recently excited and black as excitable. The simulation started from a random distribution of 54 excited cells on a 30 x 90 grid

interactions can be captured and studied in this form (Breckling et al. 2011; Makeev and Semendyaeva 2017). In these models, the transition from excitable to excited can be interpreted as a tipping event, furthermore there are parameter specifications beyond which the observable pattern transits to different characteristics. Because of its relation to ecological dispersal processes, the approach can help to understand dynamic phenomena of SPAGE dispersal the same way as conventional ecological dynamic can be modelled.

In epidemics, the according states frequently are named "susceptible, infected, recovered" and are modelled in random networks (Volz and Meyers 2007). For SPAGE the model type is of interest because it provides a way to describe long-term and large scale dispersal effects basing on a specification of strictly local interactions.

Percolation

With regard to ecological application potential, percolation theory is to some extent related to excitable media. A grid-based process is considered in which permeable and impermeable cells are randomly dispersed. In such a setting, it is statistically analysed, at which density of randomly dispersed permeable cells a continuous connection of opposite sides of the grid becomes probable (Fig. 2.6). As soon as a grid type specific density (the percolation threshold) is exceeded, percolation takes place with a high probability, otherwise not. It is remarkable that the threshold stands for a rather sharply dividing boundary which can be considered as a specific form of a tipping point. Percolation theory has various applications in physics as well as in epidemiology and ecology (see e.g. Boswell et al. 1998; Davis et al. 2008). To fire ecology, percolation theory contributed the following interesting result. In a random landscape consisting of flammable and non-flammable spatial components (local sites), fires rapidly extinguish when the overall frequency of randomly distributed highly flammable locations is below the percolation threshold. Above the percolation threshold, however, there is a high probability that local fires expand throughout the entire landscape (in a model: across the grid). At the percolation threshold itself (the tipping point), there is a singularity: the way the fire has to take across the landscape would be the longest compared to other constellations, and the intensity in terms of synchronously burning sites would be the lowest possible above extinction (Galeano Sancho 2015). In such a situation, the fire (or otherwise represented dispersal process) would persist for the longest temporal range directly before percolation would occur. For ecological applications, percolation analysis can help to decide, whether a population could sustain with regard to the particular biogeographical setting, which is an important question in an assessment of gene drive persistence.

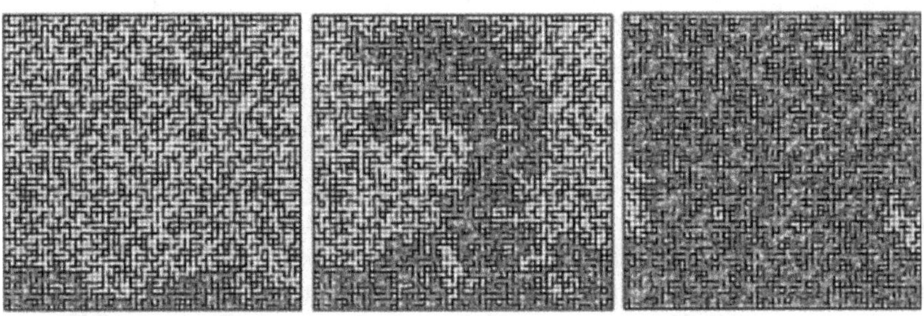

Fig. 2.6 Percolation models are used to describe dispersal processes in heterogeneous media. When an increasing rate of randomly placed local obstacles are placed across a grid, a characteristic tipping point emerges above which for almost any configuration a transition throughout the grid is possible. Three configurations are shown, one below (left), one above (right) and one close to the percolation threshold (middle). The system at the tipping point has specific characteristics. Percolation models are used e.g. in physics, in soil science but also in landscape ecology to analyse dispersal and persistence pattern of organisms in fragmented environments. For gene drive organisms the approach is highly relevant to assess overall implications of altered abilities to cope with local spatial resistance

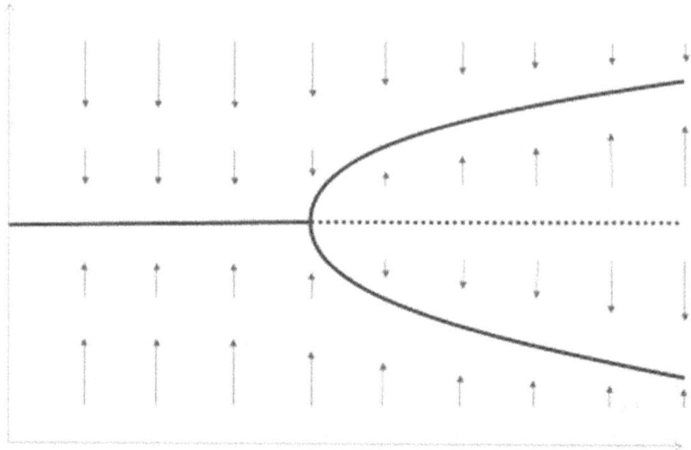

Fig. 2.7 Bifurcations occur in dynamic systems as a result of a non-linearity. A typical example is the pitchfork bifurcation. The normal form of the according differential equation is $dx/dt = r * x - x^3$. When increasing the critical parameter r, a tipping point occurs beyond which the configuration transits from a globally stable state to a configuration with two alternative stable states separated by an unstable equilibrium (dotted line). In many physical systems as well as in ecological population dynamics this type of tipping point can be observed

Threshold Effects and Phase Transitions, Bifurcations and Hopf-Bifurcation

It is a major issue in dynamic theory to analyse forms of transitions which can occur in systems in which causal structures determine the change of states over time. A typical transition occurring at a tipping point is a pitchfork bifurcation (Fig. 2.7). If a system reaches a critical state, at the bifurcation point an infinitesimally small influence can decide, to which side of a “pitchfork”, that characterizes possible stable states, the system will develop. Also for this scenario, the characterisation by Prigogine and Stengers (1984) is applicable that at the bifurcation point there is an exception in an otherwise coherent causal structure.

A bifurcation named after the Austrian mathematician E.F.F. Hopf is a transition where the stable state of a system becomes unstable and a limit cycle occurs when a critical threshold (bifurcation point or tipping point) is exceeded (Fig. 2.8). A variety of oscillatory processes in ecology can be described with a Hopf bifurcation model (e.g. Fussmann et al. 2000).

Hysteresis

For the discussion of tipping points in the given context, hysteresis is one of the significant forms of transition and plays a central role in the climate change debate. Hysteresis is a system structure with a domain in the state space where alternative

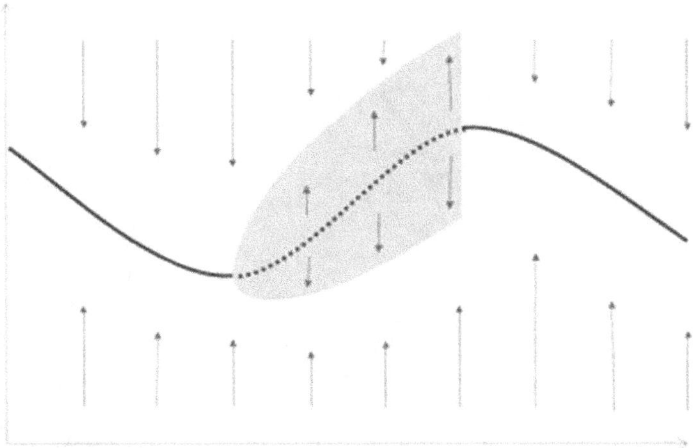

Fig. 2.8 Hopf-bifurcations occur also in population dynamics: Beyond a tipping point, a stable equilibrium becomes unstable and a limit cycle occurs. This is a transition from a stationary stable state to a stable periodic oscillation. Grey area: Domain of a critical parameter where oscillatory behaviour occurs

stable states are possible, i.e. under exactly the same external conditions the system will remain in either one of two possible states. This implies that the system description goes beyond a plain functional relationship where each input is related to only one specific output. The parameter domain in which alternative stable states occur, is usually limited. The hysteresis effect occurs, when the limiting parameter is exceeded and the system flips into the other one of two states. As a characteristic property, the transition from stable state 1 to stable state 2 happens at a different size of the critical parameter than the return from stable state 2 to stable state 1. The description of such a transition requires a three-dimensional representation (Fig. 2.9). Conceptual descriptions are given e.g. by Scheffer et al. (2001) and Gordon et al. (2008). The analysis of relations in an ecological context as well as with regard to the global climate system have brought about a large number of illustrative cases (Scheffer and van Nes 2007). For SPAGE the approach is of high interest since it provides an explanation, under which conditions a system could shift to a basically different set of states, which can be very difficult to restore and get back to previous states.

After having discussed a short survey of examples and formal constructs in which the effect of critical conditions and tipping points are captured, we now analyse potential occurrences of tipping points with regard to the new genomic constructs like SPAGE and gene drives. There are indications that some of the major forms in which tipping points occur, are highly likely to play a significant role also for an understanding of implications of a deliberate release of SPAGE.

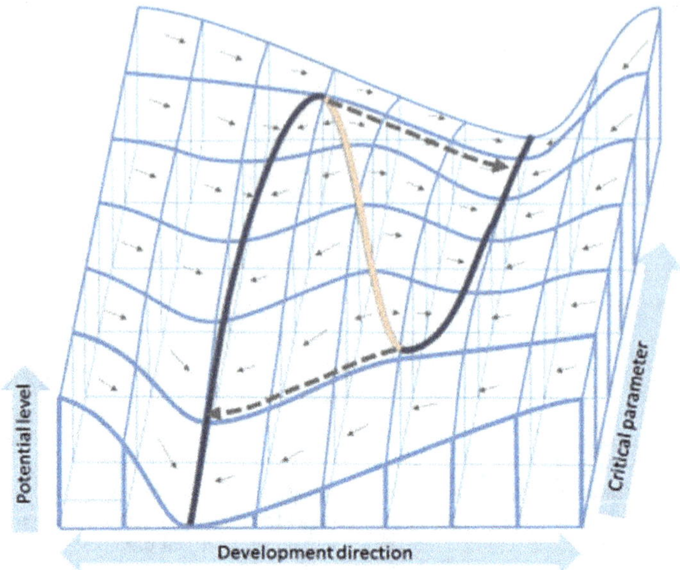

Fig. 2.9 Hysteresis can occur in an ecological context where the change of a critical parameter leads to an abrupt transition towards a new equilibrium (bold dotted arrows). There is the specific condition that the reversal occurs at a different value of the critical parameter. This comes with the implication of the existence of two alternative equilibria for a particular size range of the critical parameter. This behaviour is known for a variety of ecological systems, including shallow lakes in temperate climate. For global climate change a related configuration is also in discussion

Potential Tipping Points with SPAGE Involvement

When considering SPAGE, tipping points can be discussed on various levels. Though at first glance it may seem likely to think of molecular properties of the transgenes in first place. However, since the implications of SPAGE are necessarily embedded in a wider ecological and social context, the possibility of tipping point dynamics must be also expected in a wider spectrum of contexts. *A systematic assessment is an open and important research question, which requires substantial attention and an effort in interdisciplinary exchange.* Here, we can provide an outline and first hypotheses on tipping point dynamics and associated model forms.

A Hierarchy in Consideration Levels: Where SPAGE-Related Tipping Points Could Occur

In ecology, the assignment of interactions to specific levels of organization was established as an efficient approach for structuring and systematisation. How to do that is summed up under the term hierarchy theory (O'Neill et al. 1986). Organisation levels

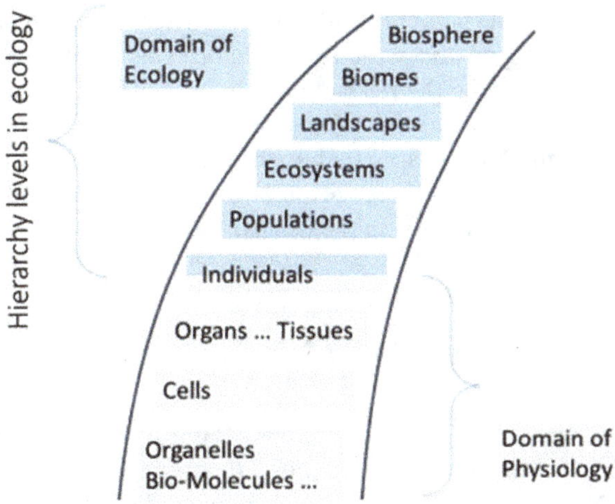

Fig. 2.10 Hierarchy theory is used to structure the topics dealt with in biology. Levels are distinguished according to emergent properties. From lower to higher levels, additional interaction types play a role which require specific forms of descriptions and which are not relevant on the lower levels. The physiological domain consists of biomolecules, cell organelles up to the level of the individual. From individual to the biosphere we have the ecological domain. To study each of the levels, requires specialised expertise

are distinguished according to specific properties that emerge on a particular level as a result of systems internal interaction and self-organisation (Fig. 2.10). In ecology, mainly the levels *organism—population—landscape—biome* and *biosphere* are distinguished. The approach is complemented with an extended view including additional levels from the molecular level up to the largest known macroscopic structures to facilitate interdisciplinary connections and views (Jantsch 1980).

Tipping Points on the Molecular and Physiological Level

The very first tipping point, proof of concept and SPAGE feasibility, can be considered as being passed. This was a transition from largely conceptual considerations (Burt 2003) to finally successful efforts on the molecular level to produce and implement an artificial genomic structure which is inherited to subsequent generations at higher rates than the natural Mendelian inheritance. From the very beginning, such a structure can be combined with other and in the end potential harmful properties on the population level (Hammond et al. 2016).

For gene drives, a next critical threshold would be a release, which means that alterations on the molecular level that lead to a manifestation on the individual level would become connected to higher levels of naturally self-organising systems, i.e. the population level and communities of organisms (ecosystems). This threshold requires, that a considerable body of regulation, risk assessment and public discourse would have to be developed and settled beforehand. This is currently not the case. Relevant questions to manage the risks, uncertainties and lack of knowledge which are created by the release of gene drive organisms are not solved—but the according scoping has been started (Oye et al. 2014). In the meantime, laboratory development under controlled conditions continues towards a differentiation and diversification of the approaches, as we described in the Chaps. 1 and 7.

Tipping Points on the Population Level

On the population level, a variety of different SPAGE constructs and species specific tipping points have to be considered. For a prospective analysis of the dynamic behaviour of SPAGE organisms detailed modelling studies would be required. The modelling set-up would have to encompass studies of individual interactions as well as large-scale spatially explicit (meta-) population representations, comparable to and expanding the modelling approaches (as laid out in Chap. 5). Phase transitions that occur when transgressing tipping points on this level can be structural shifts of systems on local, regional or potentially even global range, including extinction events. Various dynamic regime shifts, trophic network alterations including secondary higher trophic level effects, virulence changes, etc. have to be assessed. Depending on the type of SPAGE, transitions like bifurcations, limit cycles, strange attractors and hysteresis can be phenomena on this level. It has to be emphasised, that regardless of the efficiency of population level model projections, long-term predictions concerning the evolutionary fate of a specific release are likely to remain in the domain of the "unknown unknowns" (see Chaps. 1 and 9).

Tipping Points on the Ecosystem Level

The ecosystem level encompasses the overall budgets of energy and matter transformation jointly brought about by the community of organisms on the basis of the available bio-element reserves and energy basis including inputs into and outputs from the system (Jax 2016; Breckling and Koehler 2016). It is the integrating level at a specific site with sufficiently reasonable internal structural homogeneity. Comprehensive ecosystem studies are difficult, complex and therefore rare. Frequently, ecosystem investigations focus on specific ecosystem-related aspects like ecosystem services (Fischlin et al. 2007) or ecosystem integrity (Woodley et al. 1993). Both approaches are of a high importance for the impact assessment of SPAGE technologies.

Concerning systemic tipping points, an ecosystem assessment would in particular focus on key processes and key species. Ecosystem engineers (Jones et al. 1994) are species, which provide through their activity structures and conditions that largely influence the overall systems characteristics and composition of the overall community. Keystone species (Paine 1966) have a central function in the structuring of the trophic relations and thus the overall systems condition, frequently through indirect effects. If species with such a pronounced role in the ecosystem would be directly or indirectly affected through population alterations caused by a SPAGE, this might have consequences where tipping point dynamics on the ecosystem level may become apparent. An implication is, that through matter-, energy- and information-processing on the ecosystems level, minor impacts can be damped, amplified or transformed. Anticipation of these effects is notoriously difficult, but therefore not irrelevant. Knowledge in this field was accumulated in particular in conservation ecology (Schwartz et al. 2000) and in restoration ecology (Jordan et al. 1990).

Tipping Points on the Landscape and (Cross-) Regional Level

The landscape level encompasses a network of different ecosystem types (Forman 2014). Here, the spatial aspect of process organization is in the foreground. Spatially explicit approaches like grid-based models, an ecological version of cellular automata (Breckling et al. 2011) are used on this level e.g. with relation to excitable media or percolation. Furthermore, individual-based models (Łomnicki 1999; Judson 1994) with their high level-integrating potential are used to analyse the role of organism's activities across larger spatial scales in heterogeneous structures (see Chap. 5). Well known tipping points on the landscape level result from fragmentation effects (Opdam et al. 1993), and are an issue analysed in metapopulation theory (Hanski 1999) and population viability analysis (Beissinger and McCullough 2002) as far as referring to large-scale structures.

Tipping Points in an Evolutionary Context

The evolutionary context of SPAGE is probably the aspect that involves the highest degree of relevance from a perspective of risk analysis, since it deals with long-term implication on a scale and temporal extent which goes beyond direct empirical accessibility. Unfortunately, it is also the domain that allows the least precise prediction and maintains in scientific processes the highest level of irreducible uncertainty and "unknown unknowns". This is because the evolutionary process connects small-scale processes on the molecular scale, i.e. mutations, with macroscopic selection processes on the level of the whole organism through the amplification of fitness-related reproduction dynamics further on with large-scale landscape and community relations. Key issues for an understanding of evolutionary processes are considerations of a multi-level selection which acknowledge, that evolutionary change is not exclusively an interaction between genome and a particular environmentally determined fitness trait but that fitness consists of a variety of components which can be distributed and interfere across different levels. The other decisive theorem comes under the name of the "red queen hypothesis" (Van Valen 1973). This widely discussed hypothesis puts forward that fitness is neither an intrinsic property of an organism nor stationary. Instead, it is a continuously changing relation of the members of a population and their environment, which is influenced from both sides. This requires—in order to maintain a currently held position—an ongoing change. Thus, even for a stationary state, an underlying evolutionary adaptation would be required.

For evolutionary dynamics, models usually can make qualitative and probabilistic statements, which may outline expectations. Models also can give clear indication, that it is highly unlikely to predict the potential fate of a SPAGE in an evolutionary perspective. Reliable quantitative predictions in this domain can be assumed to be rare. On a meta-level we can discuss the conclusion whether tipping points of SPAGE in an evolutionary domain indicate an irreversible phase shift from known unknowns to unknown unknowns.

Tipping Point Considerations with Regard to a Social Ecological Domain of SPAGE—The Interaction of Natural and Social Processes

Systems descriptions, though mainly used in scientific processes, have been applied to describe changes and transitions in social and economic dynamics (Becker and Breckling 2011). Therefore, it is a reasonable question to study, how the availability of SPAGE may influence ongoing processes in a social ecological context (Becker and Jahn 2006) and what kind of tipping points might occur in this context. Since it is apparent, that SPAGE open up new opportunities as well as new and potentially risky and controversial developments (Oye et al. 2014; Esvelt et al. 2014), it is a useful and necessary task to identify qualitative transitions in the use of the technologies and employ the concept of tipping points to analyse and understand transitions concerning e.g. land use, health, economy, regulation, public involvement, and issues how the feedback between socio-economic activities and ecological development is operated. In the following we list a number of starting points for investigations aiming at an identification of such SPAGE related tipping points in the social ecological domain.

Tipping Points in Application and Feasibility Considerations

A first basic transition following a balance (or seesaw) paradigm was apparently passed already and this opens the space for transitions in succeeding domains. The prospects of gene drives were postulated from theoretical grounds (Burt 2003), however, their realisation turned out to be difficult if not unfeasible with means available at that time. CRISPR/Cas was originally not developed for the production of gene drives (Jinek et al. 2012). Evolved in bacteria, its natural function is to defend a bacterial cell against viral intrusion. Once CRISPR/Cas was isolated, characterized, and its mode of action understood, it became apparent that it enables various additional uses. It turned out that it could be applied also as an instrument enabling the steps which are necessary to manage the molecular obstacles in the construction and implementation of gene drives. This transition from conceptual reasoning to practical availability enabled various kind of gene drive applications (e.g. Buchman et al. (2018) for the fruit fly *Drosophila suzukii*, Gantz et al. (2015) for the mosquito *Anopheles stephensi*). What was demonstrated to be feasible now could possibly lead to further frontier crossing, requiring a discourse in additional fields like risk assessment and risk management.

Tipping Points in Risk Assessment and Management

In order to understand risks which are combined with further reaching technologies it is necessary to execute test applications under limited and controlled conditions before first applications. These experiments are set-up to study a given context so that cause-effect networks can be analysed. From an epistemological perspective, this implies specific conditions: A strict delimitation of the experiment from an unobserved surrounding, and a setting that allows to observe or detect relevant intended and potentially also unintended effects. The spatio-temporal extent of these experiments depends on the type of interaction to be studied. Extrapolation from smaller to larger scales are possible as long as re-scaling is reasonable without the interference of processes that lead to changes of the results, i.e. that hierarchy effects, phase transitions or tipping points play a role. Scaling issues are a critical point in risk assessment (Woodbury 2003). A typical example is the discussion about an adequate (and clearly limited) contribution of bio-based energy to the transition towards renewable energy (e.g. Vogt 2016). The larger the spatio-temporal extent, the more difficult and costly experimentation will become. It is apparent and generally accepted in the scientific community, that experiments that regarding their effects cannot be controlled and spatio-temporally limited are irresponsible. This implies, that there is a tipping point up to which sufficient knowledge can be gained and accountability for action can be assigned. A situation, where an experiment would be required for risk assessment that in practice could not be reversed, would represent in the terminology of Schwegler (1981) an organizational instability. In such an experiment the domain of a system's operability is left (see Fig. 2.1). A determination where such a tipping point might occur in the context of deliberation of a gene drive becomes apparent when it would be necessary to practically release a gene drive organism in order to identify implied risks. It is clear, that under controlled conditions not all parameter can be studied that play a role in the wild. In particular spreading, colonization and metapopulation dynamics are difficult to be studied because scaling effects play an important role. This makes it difficult to extrapolate results from spatio-temporally limited conditions.

Tipping Points in Regulation and Law Enforcement

Regulation and enforcement are not to be understood as a process in the same line. In particular, the entry into force of a particular law or a regulation can be considered as a tipping point, which concludes a developmental process. It is quite obvious, that regulation is not a process that follows from scientific advancement or social consensus in a direct and conclusive way. It is a complex interaction that involves social functioning and integration at large. Scientific contributions are part of the input, in addition considerations of the overall legal integrity, ethical considerations and higher level jurisdictional aspects significantly contribute.

For gene drives, it is already apparent from what we know to date, that not all questions that arise in this context are covered by existing regulations. While there are regulations to safeguard laboratory processes (in Germany issued by ZKBS 2016), there are many open questions how to facilitate a secure handling of the emerging hazard and exposure potentials beyond the lab. One of the very obvious issues which requires updated regulation is the expectable conflict with the Cartagena Protocol on Biosafety (Secretariat of the Convention on Biological Diversity 2000). As an international convention, the protocol regulates that the transboundary movement of genetically modified organisms is legal only after an informed consent of the country into which the organism is moved. This was meant to apply for organisms under controlled conditions or under cultivation. For gene drive organisms, it would be required to establish an international consent if species like olive flies, mosquitos or fruit flies would be released, even if the release would initially target only a local population. This is because the construct has a very high probability to become dispersed to other regions. A more detailed assessment of legal implications certainly would identify further open questions. The expanded technological options make it necessary to develop new and more specific and adequate regulations (see Chaps. 8 and 9).

It can be expected, that these new regulations also would require a stricter management of access to the required laboratory material to prevent abuse, as Oye et al. (2014) cautiously indicated. The requirements of an adequate handling of the technological power is currently under discussion rather than being coherently regulated (National Research Council 2015; National Academies of Sciences, Engineering, and Medicine 2016). The transition may take place between a regulatory regime that allows a trans-regional alteration of wild living species characteristics based on individual resp. partial interests or an integrated, international regulatory regime, which would, in order to be efficient, require new enforcement instruments. Most likely, a qualitative transition in trans-boundary regulation would be required to cope with the risks that can be involved by the release of self-propagating artificial genetic elements.

Tipping Points in Social Acceptance

Gene drives, as a type of SPAGE which are designed for efficient dispersal throughout natural populations, are a qualitatively new instrument in nature management. While established methods of genetic modification were justified by promises like bigger harvests and to be indispensable to feed the world with regard to the limited planetary resources, gene drive technologies promise, among others, a relief of major human diseases through elimination of vectors, elimination of significant agricultural pests, breaking of pesticide resistance developments, and elimination of invasive species (Collins 2018). Since for some of these promises, at least a technical feasibility seems credible, it requires a profound technical, ecological as well as ethical discourse.

With the experience of the debate on genetically modified crops in Europe during the last two decades it is apparent, that there seem to be tipping points also in discourse dynamics. In other upcoming technical domains risk assessment and its public perception, proponents and opponents were engaged in a constructive exchange of arguments; for the introduction of nano-materials see e.g. Filser (2019) and Wigger et al. (2015). In contrast, for the GM debate it turned out, that apparently there are at least bi-stable results possible. On the one hand arguments are used selectively as instruments to follow pre-defined business intentions rather than being primarily problem-oriented. On the other hand, unsatisfiable requests for zero risk were raised. For the discourse on SPAGE it will be an interesting research domain, to assess, how ethical considerations, scientific progress in SPAGE development, the perception of the economic potential and political influences merge either to shape public consensus or to fuel an 'us versus them' attitude. Promises, disappointment, credibility, and willingness to trust will play a significant and to-date not foreseeable role.

Discourse Outlook

SPAGE are a topic, which gains significance the more application attempts beyond laboratory conditions are proposed. The release of self-propagating artificial genetic elements which are intended to operate beyond the capability to control, withdraw or limit their proliferation in any way poses new questions. Incorporated in self-reproducing organismal entities, which are subject of evolutionary changes, of combination and interaction with an untested and untestable manifold of self-amplifying contexts, potential SPAGE application apparently requires answers to the question how to delimit domains of acceptability, how to discuss, draw and enforce lines that would be unjustifiable to be crossed. *These are typical thresholds characterized by tipping points.*

To minimise the probability that un-ethical results derive from "good intentions", it seems of high priority to debate this issue not only within the scientific community. It is required to involve not only molecular scientific expertise but also landscape ecological, geographic and ethical expertise, among others. It would be of equally high significance to facilitate a general public discourse that brings the topic to the attention of the broader public in a way that is not primarily driven by feasibility, fascination or scandalisation but sheds light to problems of knowability as well as reversibility and control.

We need to be aware that a release of SPAGE affects several tipping points. The technological power and range of SPAGE result in a range of effect potentials that cannot be holistically assessed in its implications for the biosphere. May be for the first time, humans are acquiring the potential to use the self-organisation potentials of populations to drive artificial properties into natural population including detrimental ones which have the potential to induce extinctions. It is by far not the tipping point that this would enable human induced extinctions of wild living populations for the first time. But that extinctions could be enabled by molecular tools and be lab-made and releasable is so fundamentally new, that regardless of specific application intentions, the handling of this potential requires extensive public insight, control and regulation, considerations and the highest achievable standard in ethical argumentation.

References

Andersen, T., Carstensen, J., Hernández-García, E., & Duarte, C. M. (2009). Ecological thresholds and regime shifts: Approaches to identification. *Trends in Ecology & Evolution, 24*, 49–57.

Becker, E., & Breckling, B. (2011). Border zones of ecology and systems theory. In A. Schwarz, K. Jax (Eds.), *Ecology revisited. Reflecting on concepts, advancing science* (pp. 385–403). Heidelberg.

Becker, E., & Jahn, T. (Eds.). (2006). *Soziale Ökologie: Grundzüge einer Wissenschaft von den gesellschaftlichen Naturverhältnissen*. Campus Verlag.

Beissinger, S. R., & McCullough, D. R. (2002). *Population viability analysis*. University of Chicago Press.

Bindoff, N. L., Stott, P. A., AchutaRao, K. M., Allen, M. R., Gillett, N. et al. (2014). *Detection and attribution of climate change: From global to regional*. https://pure.iiasa.ac.at/id/eprint/10552/1/Detection%20and%20attribution%20of%20climate%20change%20From%20global%20to%20regional.pdf.

Biskaborn, B. K., Smith, S. L., Noetzli, J., Matthes, H., Vieira, G., et al. (2019). Permafrost is warming at a global scale. *Nature Communications, 10*(1), 264.

Boswell, G. P., Britton, N. F., & Franks, N. R. (1998). Habitat fragmentation, percolation theory and the conservation of a keystone species. *Proceedings of the Royal Society of London B: Biological Sciences, 265*(1409), 1921–1925. https://doi.org/10.1098/rspb.1998.0521.

Breckling, B., & Koehler H., (2016). Ökosystemforschung in der zweiten Hälfte des 20. Jahrhunderts—Stoffkreisläufe und Energieflüsse in ökologischen Modellen. *Natur und Landschaft 9* (10), 410–416. https://doi.org/10.17433/9.2016.50153409.410-416.

Breckling, B., Pe'er, G., Matsinos, & Y.G. (2011). Cellular automata in ecological modelling. In *Modelling complex ecological dynamics* (pp. 105–117). Berlin, Heidelberg: Springer.

Buchman, A., Marshall, J. M., Ostrovski, D., Yang, T., & Akbari, O. S. (2018). Synthetically engineered Medea gene drive system in the worldwide crop pest *Drosophila suzukii*. *Proceedings of the National Academy of Sciences, 115*(18), 4725–4730.

Burt, A. (2003). Site-specific selfish genes as tools for the control and genetic engineering of natural populations. *Proceedings of the Royal Society of London. Series B: Biological Sciences, 270*(1518), 921–928.

Carpenter, S. R., Cole, J. J., Pace, M. L., Batt, R., Brock, W. A., Cline, T., et al. (2011). Early warnings of regime shifts: A whole-ecosystem experiment. *Science, 332*, 1079–1082.

Chandler, C., Cheney, P., Thomas, P., Trabaud, L., & Williams, D. (1983). *Fire in forestry. Volume 1. Forest fire behaviour and effects. Volume 2. Forest fire management and organization*. Wiley & Sons, Inc.

Collins, J. P. (2018). Gene drives in our future: Challenges of and opportunities for using a self-sustaining technology in pest and vector management. In *BMC Proceedings* (Vol. 12, No. 8, p. 9). BioMed Central.

Davis, S., Trapman, P., Leirs, H., Begon, M., & Heesterbeek, J. A. P. (2008). The abundance threshold for plague as a critical percolation phenomenon. *Nature, 454*(7204), 634–637. https://doi.org/10.1038/nature07053.

Dean, J. (2018). In Vox (Ed.), *Methane, climate change, and our uncertain future. Earth and Space Science News*. https://eos.org/editors-vox/methane-climate-change-and-our-uncertain-future.

deYoung, B., Barange, M., Beaugrand, G., Harris, R., Perry, R. I., et al. (2008). Regime shifts in marine ecosystems: Detection, prediction and management. *Trends in Ecology and Evolution 23*, 402–409. https://doi.org/10.1016/j.tree.2008.03.008.

Domb, C., & Green, M. S. (Eds.). (1972 ff). *Phase transitions and critical phenomena* (Vol. 1–19). San Diego et al 1971–2001, Academic Press.

Earn, D. J., Rohani, P., Bolker, B. M., & Grenfell, B. T. (2000). A simple model for complex dynamical transitions in epidemics. *Science, 287*(5453), 667–670.

Esvelt, K. M., Smidler, A. L., Catteruccia, F., & Church, G. M. (2014). Concerning RNA-guided gene drives for the alteration of wild populations. *Elife, 3*, e03401. https://doi.org/10.7554/eLife.03401.

Filbee-Dexter, K., Pittman, J., Haig, H. A., Alexander, S. M., Symons, C. C., & Burke, M. J. (2017). Ecological surprise: Concept, synthesis, and social dimensions. *Ecosphere, 8*(12), e02005. https://doi.org/10.1002/ecs2.2005.

Filser J. (2019). Pros and cons of nano-regulation and ways towards a sustainable use. In I. Eisenberger, A. Kallhoff, & C. Schwarz-Plaschg (Eds.), *Nanotechnology: Regulation and public discourse*. Rowman & Littlefield International.

Fischlin, A., Midgley, G. F., Price, J. T., Leemans, R., Gopal, B., et al. (2007). *Ecosystems, their properties, goods and services*. IPCC report AR4 climate change 2007: Impacts, adaptation, and vulnerability. https://www.ipcc.ch/report/ar4/wg2/.

Forman, R. T. (2014). *Land mosaics: The ecology of landscapes and regions 1995*, 217. Island Press.

Fussmann, G. F., Ellner, S. P., Shertzer, K. W., & Hairston, N. G. (2000). Crossing the Hopf bifurcation in a live predator-prey system. *Science, 290*(5495), 1358–1360.

Galeano Sancho, D. (2015). *Percolation theory and fire propagation in a forest*. University of Barcelona. https://diposit.ub.edu/dspace/bitstream/2445/67391/1/TFG-Galeano-Sancho-Daniel.pdf.

Gantz, V.M., Jasinskiene, N., Tatarenkova, O., Fazekas, A., Macias, et al. (2015). Highly efficient Cas9-mediated gene drive for population modification of the malaria vector mosquito Anopheles stephensi. *Proceedings of the National Academy of Sciences,112*(49), E6736–E6743.

Gatfaoui, H., & de Peretti, P. (2019).*Flickering in information spreading precedes critical transitions in financial markets. Scientific reports*, vol. 9, pp. 5671.

Goldammer, J. G. (Ed.). (2012). *Fire in the tropical biota: Ecosystem processes and global challenges* (Vol. 84). Springer Science & Business Media.

Gordon, L. J., Peterson, G. D., & Bennett, E. M. (2008). Agricultural modifications of hydrological flows create ecological surprises. *Trends in Ecology & Evolution, 23*(4), 211–219.

Grodzins, M. (1957). Metropolitan segregation. *Scientific American, 197*, 33–41.

Hallegraeff, G. M. (1993). A review of harmful algal blooms and their apparent global increase. *Phycologia, 32*(2), 79–99.

Hammond, A., Galizi, R., Kyrou, K., Simoni, A., Siniscalchi, et al. (2016). A CRISPR-Cas9 gene drive system targeting female reproduction in the malaria mosquito vector *Anopheles gambiae*. *Nature Biotechnology, 34*(1), 78.

Hanski, I. (1999). *Metapopulation ecology*. Oxford University Press.

Harley, C. D., Randall Hughes, A., Hultgren, K. M., Miner, B. G., Sorte, C. J., et al. (2006). The impacts of climate change in coastal marine systems. *Ecology Letters, 9*(2), 228–241. https://doi.org/10.1111/j.1461-0248.2005.00871.x.

Helbing, D. (2012). Modelling of socio-economic systems. In D. Helbing (Ed.), *Social self-organization. Understanding complex systems*. Berlin, Heidelberg: Springer.

Holden, A. V., Markus, M., & Othmer, H. G. (Eds.). (2013). *Nonlinear wave processes in excitable media* (Vol. 244). Springer.

Holdren, J., & Bunn, M. (2002). *Technical background nuclear basics*. Research Library. https://web.archive.org/web/20100527132103/http://www.nti.org/e_research/cnwm/overview/technical1.asp.

Holling, C. S. (1978). *Adaptive environmental assessment and management*. Wiley & Sons.

Huppert, A., Blasius, B., & Stone, L. (2002). A model of phytoplankton blooms. *The American Naturalist, 159*(2), 156–171. https://www.tau.ac.il/lifesci/zoology/members/lewi_files/documents/Huppert_AmNat_2002.pdf.

Jantsch, E. (1980). *The self-organizing universe: Scientific and human implications of the emerging paradigm of evolution*. Pergamon-Verlag.

Jax, K. (2016). Biozönose, Biotop und Ökosystem. *Natur und Landschaft, 9*(10), 417–422. https://doi.org/10.17433/9.2016.50153410.417-422.

Jiles, D. C., & Atherton, D. L. (1986). Theory of ferromagnetic hysteresis. *Journal of Magnetism and Magnetic Materials, 61*(1–2), 48–60.

Jinek, M., Chylinski, K., Fonfara, I., Hauer, M., Doudna, J. A., & Charpentier, E. (2012). A programmable dual-RNA–guided DNA endonuclease in adaptive bacterial immunity. *Science, 337*(6096), 816–821.

Johnson, E. A. (1996). *Fire and vegetation dynamics: Studies from the North American boreal forest*. Cambridge University Press.

Jones, C. G., Lawton, J. H., & Shachak, M. (1994). Organisms as ecosystem engineers. *Ecosystem management* (pp. 130–147). New York, NY: Springer.

Jordan III, W. R., Jordan, W. R., Gilpin, M. E., & Aber, J. D. (Eds.). (1990). *Restoration ecology: A synthetic approach to ecological research*. Cambridge University Press.

Judson, O. P. (1994). The rise of the individual-based model in ecology. *Trends in Ecology & Evolution, 9*(1), 9–14.

Kapral, R., & Showalter, K. (Eds.). (2012). *Chemical waves and patterns* (Vol. 10). Springer Science & Business Media.

Kim, W. C., & Mauborgne, R. (2003). Tipping point leadership. *Harvard Business Review, 81*(4), 60–69, 122.

Kvenvolden, K. A. (1988). Methane hydrates and global climate. *Global Biogeochemical Cycles, 2*(3), 221–229. https://doi.org/10.1029/GB002i003p00221.

Lenton, T. M. (2011a). Early warning of climate tipping points. *Nature Climate Change, 1*(4), 201. https://citeseerx.ist.psu.edu/viewdoc/download?doi=10.1.1.666.244&rep=rep1&type=pdf.

Lenton, T. M. (2011b). Beyond 2 °C: Redefining dangerous climate change for physical systems. *WIREs Climate Change, 2*(3), 451–461. https://doi.org/10.1002/wcc.107.

Lenton, T. M. (2012). Arctic climate tipping points. *Ambio, 41*(1), 10–22. https://doi.org/10.1007/s13280-011-0221-x.

Lenton, T. M. (2013). Environmental tipping points. *Annual Review of Environment and Resources, 38*, 1–29.

Lenton, T. M., Held, H., Kriegler, E., Hall, J. W., Lucht, W., et al. (2008). Tipping elements in the Earth's climate system. *Proceedings of the National Academy of Sciences, 105*(6), 1786–1793. https://doi.org/10.1073/pnas.0705414105.

Łomnicki, A. (1999). Individual-based models and the individual-based approach to population ecology. *Ecological Modelling, 115*(2–3), 191–198.

Makeev, A. G., & Semendyaeva, N. L. (2017). A basic lattice model of an excitable medium: Kinetic Monte Carlo simulations. *Mathematical Models and Computer Simulations, 9*(5), 636–648. https://doi.org/10.1134/S2070048217050088.

Mayergoyz, I. D. (2012). *Mathematical models of hysteresis*. Springer Science & Business Media.

Meron, E. (1992). Pattern formation in excitable media. *Physics Reports, 218*(1), 1–66.

Mosekilde, E., & Mouritsen, O. G. (Eds.). (2012). *Modelling the dynamics of biological systems: Nonlinear phenomena and pattern formation* (Vol. 65). Springer Science & Business Media.

National Academies of Sciences. (2016). *Gene drives on the horizon: Advancing science, navigating uncertainty, and aligning research with public values*. Washington, DC: The National Academies Press.

National Research Council. (2015). *Gene drive research in non-human organisms: Recommendations for responsible conduct, national academy of sciences*. Washington, DC. https://nas-sites.org/gene-drives/.

Nordhaus, W. D. (1975). *Can we control carbon dioxide?* IIASA working paper.

Notz, D. (2009). The future of ice sheets and sea ice: Between reversible retreat and unstoppable loss. *Proceedings of the National Academy of Sciences, 106*(49), 20590–20595. https://doi.org/10.1073/pnas.0902356106.

O'Neill, R. V., Deangelis, D. L., Waide, J. B., Allen, T. F., & Allen, G. E. (1986). *A hierarchical concept of ecosystems* (No. 23). Princeton University Press.

Opdam, P., van Apeldoorn, R., Schotman, A., & Kalkhoven, J. (1993). Population responses to landscape fragmentation. *Landscape ecology of a stressed environment* (pp. 147–171). Dordrecht: Springer.

Oye, K. A., Esvelt, K., Appleton, E., Catteruccia, F., Church, G., et al. (2014). Regulating gene drives. *Science, 345*(6197), 626–628.

Paine, R. T. (1966). Food web complexity and species diversity. *The American Naturalist, 100*(910), 65–75. https://doi.org/10.1086/282400.JSTOR2459379.

Parmesan, C., & Yohe, G. (2003). A globally coherent fingerprint of climate change impacts across natural systems. *Nature, 421*(6918), 37. https://doi.org/10.1038/nature01286.

Piñol, J., Beven, K., & Viegas, D. X. (2005). Modelling the effect of fire-exclusion and prescribed fire on wildfire size in Mediterranean ecosystems. *Ecological Modelling, 183*(4), 397–409.

Prigogine, I., & Stengers, I. (1984). *Order out of chaos: Man's new dialogue with nature*. New York: Verso Books.

Randalls, S. (2010). History of the 2°C climate target. *Wiley Interdisciplinary Reviews: Climate Change, 1*(4), 598–605. https://doi.org/10.1002/wcc.62.

Rijsberman, F. R., & Swart, R. J. (Eds.) (1990). *Targets and indicators of climatic change*. Report of working group II of the advisory group on greenhouse gases, Draft version. 166. Stockholm Environment Institute.

Scheffer, M., & van Nes, E. H. (2007). Shallow lakes theory revisited: Various alternative regimes driven by climate, nutrients, depth and lake size. *Hydrobiologia, 584*(1), 455–466.

Scheffer, M., Hosper, S. H., Meijer, M. L., Moss, B., & Jeppesen, E. (1993). Alternative equilibria in shallow lakes. *Trends in Ecology & Evolution, 8*(8), 275–279.

Scheffer, M., Carpenter, S., Foley, J. A., Folke, C., & Walker, B. (2001). Catastrophic shifts in ecosystems. *Nature, 413*(6856), 591.

Scheffer, M., Bascompte, J., Brock, W. A., Brovkin, V., Carpenter, S. R., et al. (2009). Early-warning signals for critical transitions. *Nature, 461*, 53–59. https://doi.org/10.1038/nature08227.

Schwartz, M. W., Brigham, C. A., Hoeksema, J. D., Lyons, K. G., Mills, M. H., & Van Mantgem, P. J. (2000). Linking biodiversity to ecosystem function: Implications for conservation ecology. *Oecologia, 122*(3), 297–305.

Schwegler, H. (1981). Structure and organization of biological systems. In G. Roth & H. Schwegler (Eds.), *Self-organizing systems* (pp. 24–38). Frankfurt, New York: Campus-Verlag.

Schwegler, H. (1985). Ökologische Stabilität. In G. Weidemann (Ed.), *Verhandlungen der Gesellschaft für Ökologie* (13, pp. 263–270). Jahrestagung Bremen : Im Auftrag der Gesellschaft.

Sornette, D. (2003). *Why stock markets crash: Critical events in complex financial systems*. Princeton University Press.

Taleb, N. N. (2007). *The black swan: The impact of the highly improbable*. New York: Random House.

Taleb, N. N., Read, R., Douady, R., Norman, J., & Bar-Yam, Y. (2014). *The precautionary principle (with application to the genetic modification of organisms). Extreme risk initiative—NYU school of engineering working paper series*. https://arXiv.org/pdf/1410.5787.pdf.

UNEP. (2000). *Cartagena protocol on biosafety to the convention on biological diversity*. https://bch.cbd.int/protocol/text/CartagenaProtocolonBiosafety.

United Nations Climate Change. (2014). *Why methane matters*. https://unfccc.int/news/new-methane-signs-underline-urgency-to-reverse-emissions.

United Nations Climate Change. (2018). *The Paris agreement*. https://unfccc.int/process-and-meetings/the-paris-agreement/the-paris-agreement.

Van Valen, L. (1973). A new evolutionary law. *Evolutionary Theory*. *1*, 1–30. https://ebme.marine.rutgers.edu/HistoryEarthSystems/HistEarthSystems_Fall2010/VanValen%201973%20Evol%20%20Theor%20.pdf.

Veraart, A. J., Faassen, E. J., Dakos, V., van Nes, E. H., Lürling, M., & Scheffer, M. (2012). Recovery rates reflect distance to a tipping point in a living system. *Nature, 481*(7381), 357. https://doi.org/10.1038/nature10723.

Vogt, M. (2016). *Ethische Fragen der Bioökonomie. Sachverständigenrat Bioökonomie Bayern/C.A.R.M.E.N.* https://www.biooekonomierat-bayern.de/dateien/Publikationen/M.Vogt_Ethische_Fragen_der_Bio%C3%B6konomie_03.2016.pdf.

Volz, E., & Meyers, L. A. (2007). Susceptible–infected–recovered epidemics in dynamic contact networks. *Proceedings of the Royal Society B: Biological Sciences, 274*(1628), 2925–2934.

von Gleich, A., & Giese, B. (2019). Resilient systems as a biomimetic guiding concept. In M. Ruth, & S. Gößling-Reisemann (Eds.), *Handbook on resilience of socio-technical systems*. Cheltenham, Northampton: Edward Elgar Publ.

Wanfeng, Y., Woodarda, R., & Sornettea, D. (2010). Diagnosis and prediction of tipping points in financial markets: Crashes and rebound. *Physics Procedia, 3*, 1641–1657.

Wigger, H., Hackmann, S., Zimmermann, T., Köser, J., Thöming, J., von Gleich, A. (2015). Influences of use activities and waste management on environmental releases of engineered nanomaterials. *Science of the Total Environment, 535*, 160–171.

Woodley, S., Kay, J., Francis, G. (Eds.). (1993). *Ecological integrity and the management of ecosystems*. Delray, Florida: St. Lucie Press.

3

Ecosystems and their Vulnerability Analysis

Carina R. Lalyer, Arnim von Gleich and Bernd Giese

Introduction

Vulnerability analysis can be seen as the counterpart to technology characterization. Technology characterisation scrutinises the intervening technology. Vulnerability analyses potentially affected systems. That may be socio-ecological, socio-technical, socio-economic or other systems. In this chapter ecological systems are in focus.

Ecosystems represent complex assemblies whose dynamics are still far from being fully understood. Modelling the behaviour of ecological systems is thus a challenging task. Extensive abstractions are required to enable the analysis of the performance of at least some of its elements under changing environmental conditions or in the presence of certain stressors. In particular, for "novel entities" (Steffen et al. 2015) like gene drives, an analysis of their possible impact on ecosystems is complicated especially because of the lack of comparative cases. In the course of a vulnerability analysis of an ecosystem affected by gene drives, assessments of potential impacts are only possible in a rather simplified form. As an ecosystem may be defined by the interactions between its biotic and abiotic elements (Chapin et al. 2011, p. 5), the vulnerability analysis of an ecosystem requires an investigation at different hierarchical levels, from species characterization to the organism's interactions with the environment including its abiotic attributes (De Lange et al. 2010). Elements of these levels as for example species that fulfil important functions like ecosystem engineers, may represent central parts of tipping point dynamics when they are affected above a certain threshold.

C. R. Lalyer (✉) · B. Giese
Institute of Safety/Security and Risk Sciences (ISR), University of Natural Resources and Life Sciences, Vienna (BOKU), Austria
e-mail: carina.lalyer@boku.ac.at

A. von Gleich
Department of Technological Design and Development, Faculty Production Engineering, University of Bremen, Bremen, Germany

To explore the relevant criteria for a vulnerability analysis of ecosystems that are confronted with a gene drive, a framework for this kind of analysis is developed in this chapter.

Ecosystem Vulnerability Analysis

The technological advances in gene drive systems opened new possibilities for the scientific community, interested parties and governments to address problems in the agricultural, conservation and public health sector. However, these developments are highly debatable. Gene drives can be used to transform, suppress or even eliminate specific species (Meghani and Kuzma 2017) that act as disease vectors, reduce biodiversity or have become agricultural pests. Currently, one of the greatest threats to biodiversity is the establishment of invasive species (Scalera 2010). Known methods to control organisms only lead to a short term suppression of the population and thus rely on continuous applications (Moro et al. 2018). However, using gene drives in wild populations requires careful considerations because the impact of the use of this new technology is far reaching and uncertain.

This uncertainty and lack of knowledge is due to research gaps concerning the biological and ecological traits of certain species and as well as knowledge gaps regarding the ecological effects of hazards and exposure after releasing gene drives (GD) into wild populations (Moro et al. 2018).

Moreover, due to many failures in the attempt to eradicate or control unwanted species,[1] a rigorous study must be performed to assess the potential impacts and risks associated with the unprecedented release of gene drive organisms (GDO). As already mentioned, vulnerability analysis is confronted with different forms of lacking knowledge, with scientific uncertainties, known unknowns and unknown unknowns. Vulnerability analysis may contribute to the reduction of any one of them, by more precise science, by modelling or by identifying tipping points.

An ecological vulnerability analysis is suggested to be applied when a specific threat for the environment is expected (De Lange et al. 2010). For example, exposure of an ecosystem to a threat could be the release of GDs into wild populations. In the light of the potential power and range of current GD systems and regarding the fact that a gradual release approach is impossible, as it is practised with common GMOs, an a priori analysis is necessary to determine how vulnerable an ecosystem is. Essential steps are the identification of its exposure and sensitivity towards such a perturbation, its internal weaknesses and tipping points and its capacity to recover or adapt following an initial perturbation (von Gleich et al. 2010; Gößling-Reisemann et al. 2013; Weißhuhn et al. 2018).

Turner et al. (2003) describe vulnerability as the "degree to which a system, subsystem or system component is likely to experience harm due to exposure to a hazard,

---

[1] E.g. for population control of rabbits (Arthur and Louzis 1988) or the introduction of the cane toad in Australia to protect the cultivation of sugar cane (Doody et al. 2015).

either a perturbation or stress/stressor." (Turner et al. 2003, p. 8074). "Vulnerability" is used in both social and natural science disciplines, where authors define it in different ways, without a consensus of its conceptualization (Füssel 2007). Newell et al. (2005 cited in Füssel 2007) even suggested that the term vulnerability is a "conceptual cluster" in interdisciplinary research.

Although ecosystem vulnerability is still a new topic, it is important to detect the potential weaknesses and adaptive capacities of an ecosystem under threat (Weißhuhn et al. 2018). Therefore, by performing such an analysis, it should be possible to estimate "the inability of an ecosystem to tolerate stressors over time and space" (Williams and Kapustka 2000, p. 1056).

According to Liverman (1990 cited in Füssel 2007, p. 155) vulnerability is related to concepts of "resilience, marginality, susceptibility, adaptability, fragility and risk" wherein Füssel (2007, p. 155) has added the concepts of "exposure, sensitivity, coping capacity [...] and robustness". When describing vulnerability, it is important to specify the system and its vulnerability to specific hazards as well as to mention the time frame (Brooks 2003 cited in Füssel 2007, p. 156).

The fundamentals of a vulnerability analysis were set by two "reduced-form models" (Turner et al. 2003, p. 8074) developed in the realm of environmental and climate assessments (White 1974 and Cutter 2001 cited in Turner et al. 2003). First, the risk-hazard model was put into place in the 1970s and 1980s in which the impact of a hazard—the risk—is defined as a function of exposure to the hazard and the "dose–response" (sensitivity) of the system exposed (Burton et al. 1978 and Kates 1985 cited in Turner et al. 2003). Due to the shortcoming of these models, like the lack of taking into account the system's abilities to amplify or reduce the impacts (Kasperson et al. 1988 and Palm 1990 cited in Turner et al. 2003; Weißhuhn et al. 2018) or the fact that the system comprises different sub-elements that react differently to the hazard (Cutter 1996 and Cutter et al. 2000, cited in Turner et al. 2003; Frazier et al. 2014), the "pressure-and-release" models were developed. In these type of models, risk is defined as a function of stress and the explicit vulnerability of the exposed system (Blaikie et al. 1994 cited in Turner et al. 2003). Although, these models mainly address social vulnerabilities in the face of natural hazards, they put forward the basis of a general vulnerability analysis (Turner et al. 2003). Subsequently, adaptive capacity has been introduced (Smit and Wandel 2006 and Engle 2011 cited in Weißhuhn et al. 2018). Opposed to the history of the concept of vulnerability in which people are susceptible to natural hazards, ecosystem vulnerability analysis follows a view where the environment is exposed to perturbations caused by humans (Birkmann and Wisner 2006; Weißhuhn et al. 2018).

Weißhuhn et al. (2018) performed a review of scientific publications focusing on environmental or ecosystem vulnerability assessments. They found out that this kind of research gained more attention starting from 2009, which denotes a rather new topic in research. Recent work aimed to create a more interdisciplinary framework defining vulnerability as a function of exposure, sensitivity and adaptive capacity (Frazier et al. 2014; Füssel 2007; Weißhuhn et al. 2018). This definition of vulnerability is the framework that is used in the present study.

An ecosystem can be considered vulnerable regarding a certain perturbation when it is highly exposed, has a high sensitivity and low adaptive capacity (Mumby et al. 2014). Thus, those ecosystems that turn out to be vulnerable need proper management strategies (Weißhuhn et al. 2018). According to Weißhuhn et al. (2018) and de Lange et al. (2005, p. 27) *vulnerability* is defined as a function of *exposure* and *sensitivity*, leading to the potential impact (Mumby et al. 2014), and *adaptive capacity* (AC). These qualities will be described in the following passages.

Exposure

Exposure describes the fact that the ecosystem is in contact with the stressor (De Lange et al. 2010). To assess exposure, Frazier et al. (2014) recommend to examine the probability of an occurrence of the disturbance or its spatial proximity, whereas, Dong et al. (2015) suggest to determine the threatened area. The probability of the exposure of an ecosystem towards a certain stressor is determined by the quality of the stressor (e.g. its persistence or pervasiveness) and the qualities of the affected systems. As it is shown in Fig. 3.1, in this study the exposure relevant qualities of the ecosystem are differentiated between qualities of the ecosystem and qualities of the wild species targeted by GDOs. For the latter qualities of the drive have to be taken into account, which influence its spread within and probably also beyond the target species population. The qualities of the drive are identified in the course of the technology characterization as described in Chap. 1. In order to assess the exposure potential of the ecosystem, as suggested by de Lange et al. (De Lange et al. 2010), a possible measurement can be the scale of exposure. In order to determine the scale, the current study has compiled the following characteristics as:

(a) Ecosystem characteristics

1. Distribution of adequate habitat conditions: it is important to know what are the conditions required for the survival of the gene drive target species;
   - The range of environmental characteristics, in which a species lives in, is called the species' ecological niche (Hutchinson 1957 cited in Chase 2011, p. 93). This niche results from evolutionary processes in which species interact with their environment and other organisms (Chase 2011). According to Chase and Leibold (2002 cited in Chase 2011, p. 93) the niche defines a species' spatial existence, biogeography, interspecies interactions, abundance and ecological role. Chase (2011) describes in his review on niche theory, that there are two components in which the definition of a niche can be divided into: first the range of biotic and abiotic characteristics that enable a species to persist in a space (Grinnell 1917; Hutchinson 1957 cited in Chase 2011, p. 94), also named the requirement component (Chase 2011) and secondly, the impact the species has on its given environment (Elton 1927 cited in Chase 2011, p. 94), known as the impact component (Chase 2011).

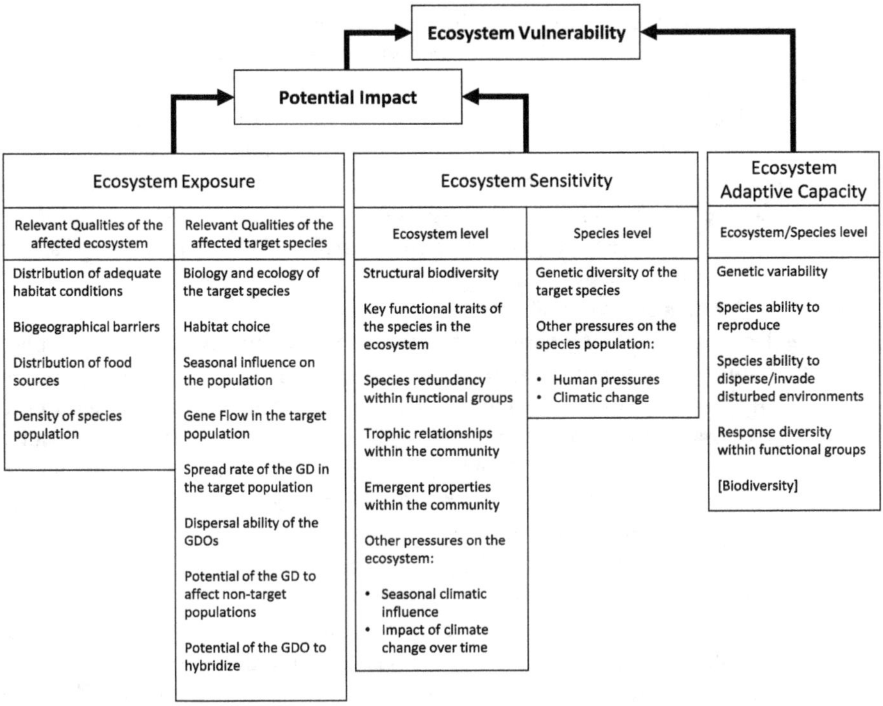

Fig. 3.1 Relevant criteria and levels of investigation for an event-based analysis of vulnerability (eVA) in adaption to Gößling-Reisemann et al. (2013) and criteria i.e. after de Lange et al. (2010), Moro et al. (2018), Weißhuhn et al. (2018) and Mumby et al. (2014)

- Huey (1991) underlines the importance of the environmental physical conditions (temperature, humidity, salinity etc.) and of the organisms' physiology to perform in a given habitat or to choose a specific habitat.

2. Biogeographical barriers (adapted from Moro et al. 2018);
   - Physical, physiological or environmental barriers influence the dispersal of a species in a certain landscape (Capinha et al. 2015).
   - According to Cox et al. (2016, pp. 91–92), the distribution of species is limited by geographical barriers that can be of different types (e.g. physical like mountains, biological like predators or climatic like temperature etc.).
   - In spite of these natural barriers that have confined species to certain locations, organisms have been able to establish themselves in places far away from their native ranges (Capinha et al. 2015). This "breakdown" of biogeographical barriers arose from human assisted dispersal through travel and trade (Capinha et al. 2015) and causes an intermixing of biota that puts the native species under additional pressure (Capinha et al. 2015; Montgomery et al. 2015). It can be expected that the locations with intensified trading relations and those that are closely located to them will suffer the greatest homogenization of biota (Capinha et al. 2015). The new

biota community would be formed by competitive generalists that will be composed of few but widespread species (McKinney and Lockwood 1999).

3. Distribution of food sources: determines where the organism might migrate;
   - One factor in selecting a habitat is the quality and locality of resources. Habitats undergo fluctuations regarding their quality over time and space (Jonzén et al. 2004). The variation is a consequence of the habitat itself or of the number of organisms using it, in relation to their density (Jonzén et al. 2004).
4. Density of species population;
   - Density-dependent habitat selection portrays the mechanisms behind habitat selection in relation to population size. The size of a population in a certain habitat is bound to density-dependent processes, due to the fact that a population can grow in size as long as the carrying capacity of the habitat allows it (Morris 2003). Fretwell and Lucas (1969 cited in Rosenzweig 1991) suggested that these processes are based on the optimal foraging and intraspecific competition principles.
   - del Monte-Luna et al. (2004, p. 485) propose a general definition of the carrying capacity which is: "the limit of growth or development of each and all hierarchical levels of biological integration, beginning with the population, and shaped by processes and interdependent relationships between finite resources and the consumers of those resources". The level of limiting resources is not constant over time, it varies according to the stochasticity of the environment, but when abstracted in models, it may be expressed as a fixed parameter (del Monte-Luna et al. 2004).

Qualities of the target species for the gene drive are also highly important to know in order to assess the exposure potential of the GDO. With regard to gene drive-specific qualities, the following list focuses on some points which are also part of the technology characterization of gene drives (see Chap. 1):

(b) Species characteristics

1. Biology and ecology of the target species (adapted from Moro et al. 2018);
2. Habitat choice: different habitats hold different living conditions (adapted from De Lange et al. 2010);
   - Morris (2003) defined habitat selection as the process through which individuals of a certain population preferentially choose to occupy or use a certain habitat based on particular variables. The selection of habitat is related to population density regulation, community interactions and the origin and maintenance of biodiversity (Morris 2003). "Habitat" according to Whittaker et al. (1973 cited in Chase 2011, p. 94) portrays the "environmental features" where a species can live. Whereas in Morris (2003, p. 2) the given definition for habitat is "a spatially bounded area, with a subset of physical and biotic conditions, within which the density

of interacting individuals, and at least one of the parameters of population growth, is different than in adjacent subsets."

3. Seasonal influence on the population (adapted from De Lange et al. 2010);
   - Seasonality produces environmental variability in terms of temperature, humidity, resource availability etc. which influences the life-history traits of organisms (Turchin 2003 cited in Taylor et al. 2013).
4. Gene flow in the target population (adapted from Moro et al. 2018);
   - Moro et al. (2018) argue that the spread of a gene throughout an ideal population is determined by random mating and whether the gene flow is high or not.
5. Spread rate (invasiveness) of the GD in the target population;
   - In gene drive systems that depend on thresholds, the spread of the GD will be determined not only by its inheritance rate but also by the number of released GDOs. If the necessary ratio of GDO to wild organisms is reached, the GD will spread and GDO numbers will further increase in the long run (Marshall and Akbari 2016).
6. Ability of dispersal: How far can the organism travel from the source population? (adapted from Moro et al. 2018);
   - This trait of the organism would also determine the gene flow between populations (Mitton 2013; Onstad and Gassmann 2014).
7. Potential of the GDO to affect non-target populations (adapted from Moro et al. 2018);
   - Gene flow facilitated by dispersal could spread the GD to non-target populations. However, Oye et al. (2014) warn that scientists have little experience with engineering natural systems for evolutionary robustness. Thus, they argue that precision drives could prevent the drive from spreading into non-intended populations, but their reliability requires further research (Oye et al. 2014). Other ways to prevent the spread to other populations than the intended one is by molecular confinement, threshold drives that will not fixate into the population at low frequencies, targeting very specific DNA sequences that are population specific or gene drives that transform the population to be sensible to a specific chemical (Esvelt et al. 2014; Marshall and Akbari 2016; Marshall and Hay 2011).
8. Potential of the GDO to hybridize (adapted from Moro et al. 2018);
   - Interspecific gene flow may happen through hybridization, introgression (David et al. 2013) or horizontal gene transfer (Werren 2011).

Sensitivity

Sensitivity of the ecosystem is the susceptibility to disturbances (Weißhuhn et al. 2018). It expresses the degree to which the system can be affected by a certain disturbance or stress. It is a quality of the affected system but also depends on the intensity of the disturbance and may change depending on the length of the exposure due to development of increased tolerance (Weißhuhn et al. 2018). De Lange et al. (2010) emphasise among other aspects the need to know the sensitivity of the community of species, their functions within the ecosystem and the trophic relationships. The following characteristics for sensitivity have been collated on the level of:

(a) Ecosystem characteristics

1. Structural biodiversity (adapted from De Lange et al. 2010), represented by species composition, population structure and number of individuals.
2. Key functional traits of species in the ecosystem: functional role of the species (adapted from De Lange et al. 2010);
   - The dynamics of ecosystems depend on the traits of organisms, their evolutionary histories and interactions in the community (Chapin et al. 2011). Therefore, it is important to understand the role of organisms in their community. Recently, the role of biodiversity in ecosystem functioning has gained popularity and appreciation (Díaz et al. 2006 in Chapin et al. 2011, p. 3).
   - Functional traits represent characteristics that allow a species to survive and reproduce and they impact its fitness. The loss or gain of species within a system can alter ecosystem processes due to the change in the species' functional traits that have great impacts on the system, namely effects on provision or limiting resources, microclimate, intraspecific or interspecific interactions and effects on disturbance regimes (Chapin et al. 2011). It is especially important in case the targeted species is or affects a key stone species (Paine 1969 cited in Bond 1994).
3. Species redundancy within functional groups (difference in sensitivity of functionally similar species) (adapted from De Lange et al. 2010);
   - The redundancy hypothesis suggests that resilience is maintained by the ability of the species to compensate through their functional role in case species are lost (Walker 1992 cited in Mitchell et al. 2000; Fonseca and Ganade 2001).
   - It is thought that the more species are in a system, the wider will be the range of conditions under which ecosystem processes can be maintained at their characteristic state (Chapin et al. 2011). Redundancy denotes diverse responses that allow ecosystem resilience to variation and change (Bengtsson et al. 2003). This is due to the theory of "diversity as insurance" (Chapin et al. 2011): Diversity ensures maintenance of functionality under extreme or novel conditions because different species do not respond in the same way to an eventual perturbation due to their evolution and life history. In other words, species diversity stabilizes ecosystem processes when e.g. annual variations happen or extreme events occur because it is unlikely that all species that perform a functional role go extinct (Walker 1995 in Chapin et al. 2011, p. 333).

4. Trophic relationships within the community (adapted from De Lange et al. 2010);
    - Energy and nutrient flow in an ecosystem is regulated by food webs (Chapin et al. 2011, p. 300). The trophic relationships that determine food webs are complex but can be narrowed down to bottom-up (e.g. productivity of plants regulate herbivore numbers) and top-down controls (e.g. predators that regulate prey population) (ibid.).
5. Emergent properties (adapted from De Lange et al. 2010);
    - According to Reuter et al. (2005), emergent properties are new qualities that form at higher integration levels and constitute more than the sum of the low-level components. The emergence concept is based in a view of nature as a hierarchical structure in which different organizational levels ranging from an individual, to community, ecosystem and landscape exist (Reuter et al. 2005). For example, dispersal characteristics of individuals of different carabid beetles have influence on the population size of these species as an emergent property (ibid.).
6. Seasonal climatic influence (adapted from De Lange et al. 2010)
7. Impact of climate change over time;
    - Can lead to additive effects. An additive effect is when the combined effects of multiple drivers are equal to the sum of the individual effects (Crain et al. 2008). Synergistic cumulative effects occur when the combined effect is greater than the sum of the individual effects (ibid.). Antagonistic cumulative effects occur when the combined effect is smaller than the sum of the individual effects (ibid.).

(b) Species characteristics

1. Genetic diversity of the species;
2. Human pressures on the species;
    - Stressors produced by humans (habitat destruction, hunting, use of pesticides) often interact and produce combined effects on biodiversity or ecosystem services[2] (Crain et al. 2008), termed additive effects.
3. Influence of climate change (adapted from De Lange et al. 2010).

Adaptive Capacity

The third step in the vulnerability assessment of ecosystems is to investigate their adaptive capacity (AC). According to Weißhuhn, adaptive capacity describes the system's ability to compensate the impacts of disturbances (Weißhuhn et al. 2018). AC is scarcely properly described for natural systems (Weißhuhn et al. 2018), however according to Folke et al. (2002) it is related to genetic diversity, biological diversity

[2]Ecosystem services are benefits that humans receive from ecological systems (Díaz et al. 2013) e.g. timber, food, biodiversity etc.

and landscape heterogeneity (Peterson et al. 1998; Carpenter et al. 2001; Bengtsson et al. 2003 cited in Folke et al. 2002).

Weißhuhn et al. (2018) suggest that AC can be measured through:

1. Genetic variability (direct relationship)
2. Species ability to reproduce (Díaz et al. 2013 cited in Weißhuhn et al. 2018)
3. Species ability to disperse in/invade into disturbed environments (Díaz et al. 2013 cited in Weißhuhn et al. 2018)
4. Response diversity within functional groups;
   - Elmqvist et al. (2003, p. 488) define response diversity as "the diversity of responses to environmental change among species that contribute to the same ecosystem function." In order to maintain desirable states of an ecosystem, after a disturbance, it is important that diverse functional groups are available to reorganize the system (Lundberg and Moberg 2003 cited in Elmqvist et al. 2003).

Mumby et al. (2014) highlight the general relevance of biodiversity for the capacity of an ecosystem to adapt. Biodiversity is considered to be the "biological diversity in a system, taking into account the genetic, species diversity and their functional roles but also ecosystem diversity in a landscape" (Chapin et al. 2011). The debate about the role of biodiversity in ecosystem resilience is ongoing.

Event-Based Analysis of Vulnerability

When potential disturbing events can already be described, it is advisable for an analysis of ecosystem vulnerability to refer to the factors exposure, sensitivity and adaptive capacity described in the previous section within the framework of an event-based vulnerability analysis (eVA). In many cases, conclusions about an expected exposure can already be drawn from the character of the disturbance. If, for example, flying insects are considered as the source of the disturbance, a comparatively high mobility and, in the case of distinct climatic tolerance, even ubiquitous distribution and thus intensive exposure can be assumed. An appropriate scheme of an eVA in adaption to Gößling-Reisemann et al. (2013) for the release of GDO is presented in Fig. 3.1.

Potential Tipping Events Caused by GDO

In ecosystems, tipping points can be defined for different dimensions of potential outcomes. Besides a loss of biodiversity by a population reduction, ecosystem functions and services might as well be affected. Moreover, following the disappearance of a population or a species, functional shifts within the niche of the suppressed species may occur.

In the case of population suppression, tipping points are already reached when the size of the target population is decreased below a certain threshold at which the population becomes instable and potentially disappears. This also applies to non-target species when they are affected due to an interspecific spread of the GD or indirect effects that are caused by the suppression of the target population. An overview on the variety of tipping events that may follow a population suppression is given in Fig. 3.2.

Among the potential impacts of the release of a suppression drive a number of effects represent tipping points. With regard to controllability, a tipping point is already reached when a GD appears in a non-target population. For an application as suppression drive against an invasive species this would mean that it cannot be guaranteed that the drive remains limited to invasive populations. If populations of the target species in their native habitat are also concerned, a suppression might have serious consequences for the respective ecosystem. The worst case would be the eradication of a non-target population, an event that marks a further tipping point in this direction. In general, each local extinction of a target species can be regarded as a tipping event—not least with regard to its potential irreversibility. The final tipping point would then be the global eradication of the target species.

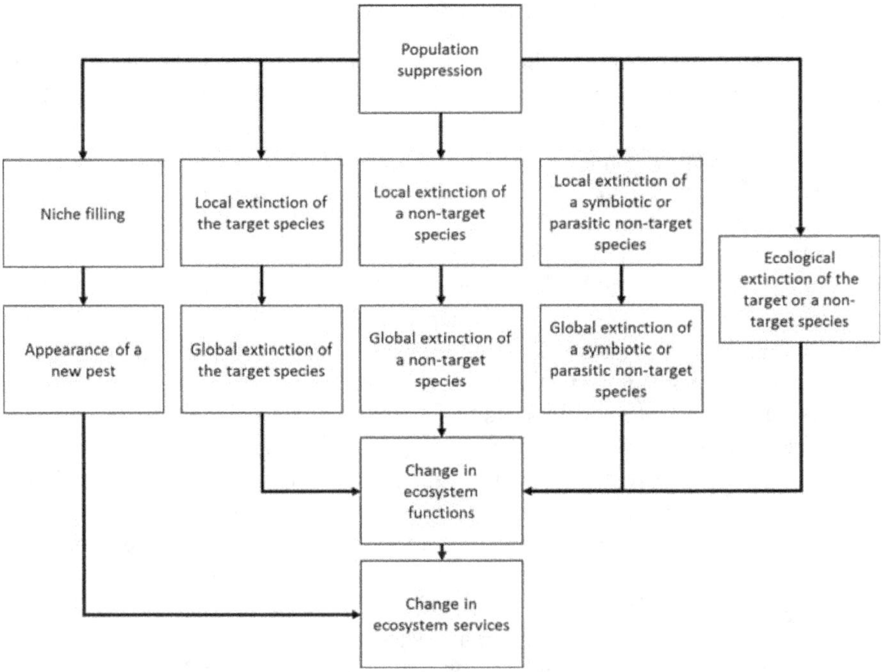

Fig. 3.2 Possible tipping points following a population suppression

Besides an unintentionally strong reduction of the target population with the outcomes as described above, the transfer of a GD from the target species to a non-target species marks an additional dimension of tipping events. Similar to the cases discussed above it may lead to either the eradication of a non-target species population or—as the worst case—to a global extinction of the non-target species.

The extinction of a species on the global or local level may cause different effects within the ecosystem. Adjustments in the abundance and population structure of other species are more distinct if the species that becomes extinct has an important role and therefore a strong interaction with other species as e.g., predator, prey or competitor (Estes et al. 1989). Impacts manifest not only because of the disappearance of a species. Estes et al. recognized that besides global and local extinction also the reduction of a species below a certain level can impede a significant interaction with other species (Estes et al. 1989). They coined the term "ecological extinction" for this class of impairment. Accompanying changes in ecosystem functions and moreover, in ecosystem services would therefore represent tipping points.

With regard to socio-economic systems, the appearance of a new exotic pest or the appearance of a secondary pest due to niche filling after the suppression or eradication of a species denotes a further tipping event.

Structural Analysis of Vulnerability

An event based analysis of the vulnerability of ecosystems, for which the case studies in this volume exemplarily provide preparatory work, will focus on the qualities described above for the characterization of a potential exposure of the system, its sensitivity and adaptive capacity. Still more difficult is the situation when we try not only to consider disturbances that are already known but also unknown perturbations, so-called ecological surprises (Filbee-Dexter et al. 2017). In this case not an event based vulnerability analysis (eVA) but a structural vulnerability analysis (sVA) of the system is the method of choice (Gößling-Reisemann et al. 2013). The structural vulnerability analysis asks at which elements or relations the system will surrender, when it comes under pressure. The distinction between eVA and sVA additionally meets the requirements of the fact that an ecosystem is not only threatened by external stressors but also by internal stressors, weak points and tipping points. Whereas an eVA is oriented along an analysis of exposure, sensitivity and adaptive capacity of the potentially affected system, the structural analysis excludes exposure and sensitivity and instead focuses on critical elements within the system that primarily account for its vulnerability. With the term critical elements an additional aspect comes into view regarding the ecosystem services of the system. Critical elements refer to elements that are essential (of high value) for the society that is dependent on the systems services (e. g. feed or food, healthy air and water etc.). Independent of any external perturbations, the specific condition and structure of potentially affected ecosystems yields important information on their general vulnerability. A structural analysis of vulnerability (sVA) lays its focus on the structure and the adaptive capacity of the system, on its capacities to cope with unknown external stressors, as well as internal weak points, tipping points and critical elements and thus on the capability to maintain system services.

Resilience of Ecosystems

Investigations of inevitable competencies, construction elements and construction principles of systems for their resilience yielded adaption and resistance (or robustness) as necessary abilities and (e.g.) redundancy, diversity and self-organization as important construction principles (von Gleich and Giese 2019). Although different authors define adaptive capacity as either "potential of recovery" or "resilience" (Weißhuhn et al. 2018), both of the concepts are being characterized by the ecosystem's biotic elements (Oliver et al. 2015; Thrush et al. 2009; Weißhuhn et al. 2018). But apart from a confusing mixture of the concept of resilience with the capacity to adapt (as an element of vulnerability analysis), the full potential of resilience can only be tapped when both terms are applied separately.

Through resilience, ecosystems maintain relatively stable functionality over long periods of time despite fluctuations in the environment. Holling introduced resilience in ecosystem theory as the capacity to "absorb changes of state variables, driving variables, and parameters, and still persist." (Holling 1973, p. 17). Thereby, resilience determines the persistence of systems—or their extinction. According to Thrush et al. (2009), resilience is the potential for recovery from disturbance (Pimm 1991 cited in Thrush et al. 2009). Holling later referred to this definition of resilience as engineering resilience (Holling 1996). An indicator for engineering resilience is seen in the "duration of the recovery phase" (Weißhuhn et al. 2018). Mitchell et al. (2000) state that an ecosystem returns faster in time to equilibrium after a perturbation when its resilience is high. The second definition is that of the ecological resilience "a variable that represents the movement of an ecosystem within and between stability domains" (Thrush et al. 2009, p. 3209, see also Gunderson 2000; Ludwig et al. 1997; Holling 1996). Walker et al. define ecological resilience as "the capacity of a system to absorb disturbance and reorganize while undergoing change so as to still retain essentially the same function, structure, identity, and feedbacks" (Walker et al. 2004, p. 5). According to Thrush et al. (2009) engineering resilience can be used to measure resilience empirically while ecological resilience requires measurement over a long time period. Besides the differentiation of engineering and ecological resilience, a transition in the notion of resilience occurred in that it was once focused on the conservation of structure integrity and is now also considering reorganization of the affected system (Oliver et al. 2015).

Ecosystems are resilient to regimes of natural variations like daily, seasonal or annual cycles and to extreme events that occurred already throughout their evolutionary history. Positive and negative feedbacks are of high importance to maintain the internal dynamics of an ecosystem (Hanski et al. 2001; Chapin et al. 2011). Negative feedbacks are the ones that stabilize the system and confer resilience (Chapin et al. 2011).

Oliver et al. (2015) suggest the same characteristics as being descriptive for ecosystem resistance or recovery. They differentiate between the following:

(a) The level of species:

- Sensitivity towards change
- Rate of population increase
- Adaptive phenotypic plasticity
- Genetic variability and dispersion (portfolio effect)
- No growth suppression in the case of low population density (allee effects).

(b) The level of the communities of species through:

- Correlation between the traits affected by change and those traits which are important for ecosystem functions
- Functional redundancy (combined with varying responses to environmental perturbations)
- Highly connected nested networks of species with generalized interactions versus networks with strong specialized interactions.

(c) The level of landscapes:

- Level of heterogeneity in the local environment
- Landscape level functional connectivity
- Possibility of alternative stable states
- Spaciousness which generally promises resource wealth.

In addition, Thrush et al. (2009) stress that resilience is being influenced by:

- Metacommunity structure (pattern of spatial dimensions of specific ecological communities)
- Community connectivity
- β-diversity (quantifies the difference between total species diversity of a region [γ-diversity] and local species diversity [α-diversity] and reflects the species turnover between the different locations in a region).

Moreover, the ecological memory is an important capacity of ecosystems to achieve resilience (Walter et al. 2013 cited in Weißhuhn et al. 2018). The ecological memory shapes how ecosystems react in the face of disturbance regimes and is defined as the "ability of the past to influence the present trajectory of the system" (Peterson 2002 cited in Hughes et al. 2019, p. 40). Depending on the ecological memory which can also be manifested through the species life-history traits or different biotic or abiotic structures like presence of certain species, the ecosystem can be resilient or vulnerable in the face of disturbances (Johnstone et al. 2016).

However, although this listing may be tempting to derive resilience from a mere description of the system under study, Thrush et al. (2009) argues that empirical studies are not sufficient to measure resilience. Instead, there is a need to develop models and identify the positive feedbacks that would drive systems to change.

Regime Shifts and Resilience

The ecological definition of resilience states that a variable of the ecosystem can move "within and between stability domains" (Ludwig et al. 1997 and Gunderson 2000 cited in Thrush et al. 2009). It has been proposed that there is not one stable equilibrium in which the ecosystem can be (Chapin et al. 2011, p. 7) but rather that systems may have alternative stable states reached by abrupt shifts (Oliver et al. 2015) that are determined by large disturbances (Beisner et al. 2003). Alternative stable states have been proposed for the first time in the late 1960s by (Lewontin 1969 cited in Beisner et al. 2003) in reference to communities of organisms (Beisner et al. 2003). According to Beisner et al. (2003), the concept of alternative stable states is being used in ecology in two ways: first, it refers to stability in population ecology (Lewontin 1969 and Sutherland 1974 cited in Beisner et al. 2003). In models of population ecology, the environment is in a fixed state where the biotic community has "different stable configurations" and secondly, the ecosystem perspective focuses on the effects of environmental change (May 1977 cited in Beisner et al. 2003). The variables and characteristics of the communities or ecosystems will persist in different possible arrangements, contributing to an alternate stable state (Beisner et al. 2003).

Therefore, if an ecosystem is resilient, it may enter into an alternative stable state, but if resilience is reduced by for example limiting species redundancy, reducing response diversity or human made pressures, the ecosystem may abruptly shift to a less desirable state (Folke et al. 2004), due to the fact that it may have reached a tipping point. The scientific community still debates when a different state can be named alternative but it is agreed that identification of critical variables and how they are affected requires a thorough understanding of species interactions and feedbacks between the biotic and abiotic elements of the ecosystem (Beisner et al. 2003). Thrush et al. (2009) suggest the following indicators for implications of disturbances or when there is a risk of regime shift (see also Chap. 2 on tipping points for case specific as well as more general indications):

- Communities are homogenising
- The complexities of food webs decrease
- Diversity within functional groups decreases
- Biogenic habitat structure decreases
- Size of organisms decreases
- Decrease in abundance of key species or key functional groups
- Changes in productivity
- Changes in recruitment and juvenile mortality
- Changes in the timing of events which lead to a decoupling of processes.

In order to further unravel the concepts of vulnerability and resilience and at the same time exploit the potential of the latter, it is recommended to shift the use of the resilience concept from an analytical category (related to ecosystem stability or maintenance of ecosystem services) to a guiding concept of the design of resilient socio-ecological systems. In von Gleich and Giese (2019) corresponding construction principles are listed.

Summary

The prospective analysis of the vulnerability of ecosystems is an extremely demanding task considering the necessary knowledge about the elements of an ecosystem and their interaction. According to the current state of research, the vulnerability of an ecosystem is dependent on three main criteria: The magnitude of exposure, the system's sensitivity and its adaptive capacity. Thus, an ecosystem can be considered vulnerable when the magnitude of its exposure is high, its sensitivity is high and its adaptive capacity is low. For each of the three criteria, factors could be identified that significantly influence the impact of a stressor (such as a GDO) on an ecosystem.

In the context of an event-based vulnerability analysis, the categories exposure, sensitivity and adaptive capacity are used to assess the vulnerability of an ecosystem. However, the large number of potentially relevant factors in question can only be partially assessed within the framework of an event based vulnerability analysis.

If an event-based analysis is hindered by lack of knowledge about potential stressors and their possible impacts, a structural analysis of vulnerability can give first hints on weak points, tipping points and critical elements and the general susceptibility of the potentially affected ecosystem to stress. For a structural analysis it is assumed that all elements and relations of the system are subject to stress. The focus is then on the question which of these elements or relations would most likely give way in case of perturbance. Subsequently, it has to be investigated whether a possibility can be identified that a specific stressor like a GDO is able to negatively influence the identified critical elements, relations or weak points.

However, beyond the described indicators that should help to assess the potential vulnerability of an ecosystem, the system's fate is determined by its resilience, a capacity that not only comprises a conservation of structure integrity because it includes the capability for a reorganization of the affected system as well. But reorganization may thereby lead to alternative stable states which may have a less desirable character for humankind.

In the following chapters on case studies some elements of a prospective vulnerability analysis are performed, by the ecological characterisation of olive flies and oilseed rape (also regarding gene flow within and between species) and by the modelling of dispersal and invasiveness. However, these studies can only be seen as preliminary approaches whose value is not least an identification of the knowledge gaps that have to be filled before a more comprehensive analysis of vulnerability is possible.

References

Arthur, C. P., & Louzis, C. (1988). A review of myxomatosis among rabbits in France. *Scientific and Technical Review of the Office International des Epizooties*, *7*, 959–976.

Beisner, B., Haydon, D., & Cuddington, K. (2003). Alternative stable states in ecology. *Frontiers in Ecology and the Environment, 1*, 376–382. https://doi.org/10.1890/1540-9295(2003)001[0376:ASSIE]2.0.CO;2.

Bengtsson, J., Angelstam, P., Elmqvist, T., Emanuelsson, U., Folke, C., Ihse, M., Moberg, F., & Nyström, M. (2003). Reserves, resilience and dynamic landscapes. *AMBIO: A Journal of the Human Environment, 32*, 389–396.

Birkmann, J., & Wisner, B. (2006). Measuring the un-measurable: The challenge of vulnerability; report of the second meeting of the UNU-EHS expert working group on measuring vulnerability, 12–14 October 2005, Bonn, Germany, Studies of the University: Research, Counsel, Education. Bonn: UNU-EHS.

Blaikie, P., Cannon, T., Davies, I., & Wisner, B. (1994). At risk. London: Routledge.

Bond, W. J. (1994). Keystone species. In E.-D. Schulze & H. A. Mooney (Eds.), *Biodiversity and Ecosystem Function. Praktische Zahnmedizin Odonto-Stomatologie Pratique Practical Dental Medicine (Geology)* (vol. 99, pp. 237–253). Berlin, Heidelberg: Springer. https://doi.org/10.1007/978-3-642-58001-7_11

Brooks, N. (2003). Vulnerability, risk and adaptation: A conceptual framework (No. Working Paper No. 38). Tyndall Centre for Climate Change Research.

Burton, I., Kates, R. W., & White, G. F. (1978). The environment as hazard. Oxford: Oxford University.

Capinha, C., Essl, F., Seebens, H., Moser, D., & Pereira, H. M. (2015). The dispersal of alien species redefines biogeography in the Anthropocene. *Science, 348*, 1248–1251. https://doi.org/10.1126/science.aaa8913.

Carpenter, S. R., Walker, B., Anderies, J., & Abel, N. (2001). From metaphor to measurement: Resilience of what to what? *Ecosystems, 4*, 765–781. https://doi.org/10.1007/s10021-001-0045-9.

Chapin, F. S., Matson, P. A., & Vitousek, P. M. (2011). Principles of terrestrial ecosystem ecology (2nd ed.). New York: Springer.

Chase, J. M. (2011). Ecological Niche theory. In S. M. Scheiner & M. R. Willig (Eds.), *The theory of ecology*. University of Chicago Press.

Chase, J. M., & Leibold, M. A. (2002). Spatial scale dictates the productivity–biodiversity relationship. *Nature, 416*, 427–430. https://doi.org/10.1038/416427a.

Cox, C. B., Moore, P. D., & Ladle, R. J. (2016). Biogeography: An ecological and evolutionary approach (9th ed.). Wiley Inc.

Crain, C. M., Kroeker, K., & Halpern, B. S. (2008). Interactive and cumulative effects of multiple human stressors in marine systems. *Ecology Letters, 11*, 1304–1315. https://doi.org/10.1111/j.1461-0248.2008.01253.x.

Cutter, S. L. (1996). Vulnerability to environmental hazards. *Progress in Human Geography, 20*, 529–539. https://doi.org/10.1177/030913259602000407.

Cutter, S. (2001). American hazardscapes. Washington, D. C.

Cutter, S. L., Mitchell, J. T., & Scott, M. S. (2000). Revealing the vulnerability of people and places: A case study of georgetown county, South Carolina. *Annals of the Association of American Geographers, 90*, 713–737. https://doi.org/10.1111/0004-5608.00219.

David, A. S., Kaser, J. M., Morey, A. C., Roth, A. M., & Andow, D. A. (2013). Release of genetically engineered insects: A framework to identify potential ecological effects. *Ecology and Evolution, 3*, 4000–4015. https://doi.org/10.1002/ece3.737.

De Lange, H. J., Van Der Pol, J. J.., Lahr, J., & Faber, J. H. (2005). A conceptual approach to assess impact of environmental stressors. In *Ecological vulnerability in wildlife*. Wageningen: Alterra.

De Lange, H. J., Sala, S., Vighi, M., & Faber, J. H. (2010). Ecological vulnerability in risk assessment—A review and perspectives. *Science of the Total Environment, 408*, 3871–3879. https://doi.org/10.1016/j.scitotenv.2009.11.009.

del Monte-Luna, P., Brook, B. W., Zetina-Rejon, M. J., & Escalona, V. H. (2004). The carrying capacity of ecosystems. *Global Ecology and Biogeography, 13*, 485–495. https://doi.org/10.1111/j.1466-822X.2004.00131.x.

Díaz, S., Fargione, J., Chapin, F. S., & Tilman, D. (2006). Biodiversity loss threatens human well-being. *PLoS Biology, 4*, e277. https://doi.org/10.1371/journal.pbio.0040277.

Díaz, S., Purvis, A., Cornelissen, J. H. C., Mace, G. M., Donoghue, M. J., Ewers, R. M., et al. (2013). Functional traits, the phylogeny of function, and ecosystem service vulnerability. *Ecology and Evolution, 3*, 2958–2975. https://doi.org/10.1002/ece3.601.

Dong, Z., Pan, Z., An, P., Wang, L., Zhang, J., He, D., et al. (2015). A novel method for quantitatively evaluating agricultural vulnerability to climate change. *Ecological Indicators, 48*, 49–54. https://doi.org/10.1016/j.ecolind.2014.07.032.

Doody, J. S., Soanes, R., Castellano, C. M., Rhind, D., Green, B., McHenry, C. R., et al. (2015). Invasive toads shift predator–prey densities in animal communities by removing top predators. *Ecology, 96*, 2544–2554. https://doi.org/10.1890/14-1332.1.

Elmqvist, T., Folke, C., Nyström, M., Peterson, G., Bengtsson, J., Walker, B., et al. (2003). Response diversity, ecosystem change, and resilience. *Frontiers in Ecology and the Environment, 1*, 488–494. https://doi.org/10.1890/1540-9295(2003)001[0488:RDECAR]2.0.CO;2.

Elton, C. S. (1927). Animal ecology. London: Sidgwick and Jackson.

Engle, N. L. (2011). Adaptive capacity and its assessment. *Global Environmental Change, 21*, 647–656. https://doi.org/10.1016/j.gloenvcha.2011.01.019.

Estes, J. A., Duggins, D. O., & Rathbun, G. B. (1989). The ecology of extinctions in kelp forest communities. *Conservation Biology, 3*, 252–264. https://doi.org/10.1111/j.1523-1739.1989.tb00085.x.

Esvelt, K. M., Smidler, A. L., Catteruccia, F., & Church, G. M. (2014). Concerning RNA-guided gene drives for the alteration of wild populations. *eLife, 3*, e03401. https://doi.org/10.7554/eLife.03401

Filbee-Dexter, K., Pittman, J., Haig, H. A., Alexander, S. M., Symons, C. C., & Burke, M. J. (2017). Ecological surprise: Concept, synthesis, and social dimensions. *Ecosphere, 8*, e02005. https://doi.org/10.1002/ecs2.2005.

Folke, C., Carpenter, S., Elmqvist, T., Gunderson, L., Walker, B., Bengtsson, J., Berkes, F., Colding, J., Danell, K., Falkenmark, M., Gordon, L., Kasperson, R., Kinzig, A., Levin, S., Mäler, K.-G., Ohlsson, L., Olsson, P., Ostrom, E., Reid, W., Rockström, J., Savenije, H., & Svedin, U. (2002). Building adaptive capacity in a world of transformations. Series on Sustainable Development. Stockholm: Environmental Advisory Council Ministry of the Environment.

Folke, C., Carpenter, S., Walker, B., Scheffer, M., Elmqvist, T., Gunderson, L., et al. (2004). Regime shifts, resilience, and biodiversity in ecosystem management. *Annual Review of Ecology, Evolution, and Systematics, 35*, 557–581.

Fonseca, C. R., & Ganade, G. (2001). Species functional redundancy, random extinctions and the stability of ecosystems. *Journal of Ecology, 89*, 118–125. https://doi.org/10.1046/j.1365-2745.2001.00528.x.

Frazier, T. G., Thompson, C. M., & Dezzani, R. J. (2014). A framework for the development of the SERV model: A spatially explicit resilience-vulnerability model. *Applied Geography, 51*, 158–172. https://doi.org/10.1016/j.apgeog.2014.04.004.

Fretwell, D. F., & Lucas, H. L. (1969). On territorial behavior and other factors influencing habitat distribution in birds. *Acta Biotheoretica, 19*, 16–36.

Füssel, H.-M. (2007). Vulnerability: A generally applicable conceptual framework for climate change research. *Global Environmental Change, 17*, 155–167. https://doi.org/10.1016/j.gloenvcha.2006.05.002.

Gößling-Reisemann, S., Wachsmuth, J., Stührmann, S., & von Gleich, A. (2013). Climate change and structural vulnerability of a metropolitan energy system: The case of Bremen-Oldenburg in Northwest Germany. *Journal of Industrial Ecology, 17*, 846–858. https://doi.org/10.1111/jiec.12061.

Grinnell, J. (1917). The niche-relationships of the California thrasher. *The Auk, 34*, 427–433.

Gunderson, L. H. (2000). Ecological resilience—In theory and application. *Annual Review of Ecology and Systematics, 31*, 425–439. https://doi.org/10.1146/annurev.ecolsys.31.1.425.

Hanski, I., Henttonen, H., Korpimäki, E., Oksanen, L., & Turchin, P. (2001). Small-rodent dynamics and predation. *Ecology, 82*, 1505–1520. https://doi.org/10.1890/0012-9658(2001)082[1505:SRDAP]2.0.CO;2.

Holling, C. S. (1973). Resilience and stability of ecological systems. *Annual Review of Ecology and Systematics, 4*, 1–23. https://doi.org/10.1146/annurev.es.04.110173.000245.

Holling, C. S. (1996). Engineering resilience versus ecological resilience. In P. E. Schulze (Ed.), Engineering within ecological constraints (pp. 31–43). Washington D.C.: National Academy Press.

Huey, R. B. (1991). Physiological consequences of habitat selection. *The American Naturalist, 137*, S91–S115. https://doi.org/10.1086/285141.

Hughes, T. P., Kerry, J. T., Connolly, S. R., Baird, A. H., Eakin, C. M., Heron, S. F., et al. (2019). Ecological memory modifies the cumulative impact of recurrent climate extremes. *Nature Climate Change, 9*, 40–43. https://doi.org/10.1038/s41558-018-0351-2.

Hutchinson, E. (1957). Concluding remarks. Cold Spring Harbor Symposia on Quantitative Biology, *22*, 415–427. https://doi.org/10.1101/SQB.1957.022.01.039.

Johnstone, J. F., Allen, C. D., Franklin, J. F., Frelich, L. E., Harvey, B. J., Higuera, P. E., et al. (2016). Changing disturbance regimes, ecological memory, and forest resilience. *Frontiers in Ecology and the Environment, 14*, 369–378. https://doi.org/10.1002/fee.1311.

Jonzén, N., Wilcox, C., & Possingham, H. P. (2004). Habitat selection and population regulation in temporally fluctuating environments. *The American Naturalist, 164*, E103–E114. https://doi.org/10.1086/424532.

Kasperson, R. E., Renn, O., Slovic, P., Brown, H. S., Emel, J., Goble, R., et al. (1988). The social amplification of risk: a conceptual framework. *Risk Analysis, 8*, 177–187. https://doi.org/10.1111/j.1539-6924.1988.tb01168.x.

Kates, R. W. (1985). Climate impact assessment. New York: Wiley.

Lewontin, R. C. (1969). The meaning of stability. *Symposia in Biology, 22*, 13–23.

Liverman, D. M. (1990). Vulnerability to global environmental change. In R. E. Kasperson, K. Dow, D. Golding, J. X. Kasperson (Eds.), Understanding global environmental change: The contributions of risk analysis and management. Worcester, MA: Clark University.

Ludwig, D., Walker, B., & Holling, C. S. (1997). Sustainability, stability, and resilience. *Conservation Ecology, 1*. https://doi.org/10.5751/ES-00012-010107.

Lundberg, J., & Moberg, F. (2003). Mobile link organisms and ecosystem functioning: Implications for ecosystem resilience and management. *Ecosystems, 6*, 0087–0098. https://doi.org/10.1007/s10021-002-0150-4.

Marshall, J. M., Akbari, O. S. (2016). Gene drive strategies for population replacement. In *Genetic control of malaria and dengue* (pp. 169–200). Elsevier. https://doi.org/10.1016/B978-0-12-800246-9.00009-0.

Marshall, J. M., & Hay, B. A. (2011). Confinement of gene drive systems to local populations: A comparative analysis. *Journal of Theoretical Biology, 294*, 153–171. https://doi.org/10.1016/j.jtbi.2011.10.032.

May, R. M. (1977). Thresholds and breakpoints in ecosystems with a multiplicity of stable states. *Nature, 269*, 471–477. https://doi.org/10.1038/269471a0.

McKinney, M. L., & Lockwood, J. L. (1999). Biotic homogenization: A few winners replacing many losers in the next mass extinction. *Trends in Ecology & Evolution, 14*, 450–453. https://doi.org/10.1016/S0169-5347(99)01679-1.

Meghani, Z., & Kuzma, J. (2017). Regulating animals with gene drive systems: Lessons from the regulatory assessment of a genetically engineered mosquito. *Journal of Responsible Innovation, 5*, S203–S222.

Mitchell, R. J., Auld, M. H. D., Le Duc, M. G., & Robert, M. H. (2000). Ecosystem stability and resilience: a review of their relevance for the conservation management of lowland heaths. *Perspectives in Plant Ecology, Evolution and Systematics, 3*, 142–160. https://doi.org/10.1078/1433-8319-00009.

Mitton, J. B. (2013). Gene flow. *Brenner's Encyclopedia of genetics* (pp. 192–196). Elsevier. https://doi.org/10.1016/B978-0-12-374984-0.00589-1.

Montgomery, W. I., Montgomery, S. S. J., & Reid, N. (2015). Invasive alien species disrupt spatial and temporal ecology and threaten extinction in an insular, small mammal community. *Biological Invasions, 17*, 179–189. https://doi.org/10.1007/s10530-014-0717-y.

Moro, D., Byrne, M., Kennedy, M., Campbell, S., & Tizard, M. (2018). Identifying knowledge gaps for gene drive research to control invasive animal species: The next CRISPR step. *Global Ecology and Conservation, 13*, e00363. https://doi.org/10.1016/j.gecco.2017.e00363.

Morris, D. W. (2003). Toward an ecological synthesis: A case for habitat selection. *Oecologia, 136*, 1–13. https://doi.org/10.1007/s00442-003-1241-4.

Mumby, P. J., Chollett, I., Bozec, Y.-M., & Wolff, N. H. (2014). Ecological resilience, robustness and vulnerability: How do these concepts benefit ecosystem management? *Current Opinion in Environmental Sustainability, 7*, 22–27. https://doi.org/10.1016/j.cosust.2013.11.021.

Newell, B., Crumley, C. L., Hassan, N., Lambin, E. F., Pahl-Wostl, C., Underdal, A., et al. (2005). A conceptual template for integrative human–environment research. *Global Environmental Change, 15*, 299–307. https://doi.org/10.1016/j.gloenvcha.2005.06.003.

Oliver, T. H., Heard, M. S., Isaac, N. J. B., Roy, D. B., Procter, D., Eigenbrod, F., et al. (2015). Biodiversity and resilience of ecosystem functions. *Trends in Ecology & Evolution, 30*, 673–684.

Onstad, D. W., & Gassmann, A. J. (2014). Concepts and complexities of population genetics. In *Insect resistance management* (pp. 149–183). Elsevier. https://doi.org/10.1016/B978-0-12-396955-2.00005-9.

Oye, K. A., Esvelt, K., Appleton, E., Catteruccia, F., Church, G., Kuiken, T., et al. (2014). Regulating gene drives. *Science, 345*, 626–628. https://doi.org/10.1126/science.1254287.

Palm, R. I. (1990). Natural hazards: An integrative framework for research and planning. *Progress in Human Geography, 16*, 142–144. https://doi.org/10.1177/030913259201600128.

Peterson, G. D. (2002). Contagious disturbance, ecological memory, and the emergence of landscape pattern. *Ecosystems, 5*, 329–338. https://doi.org/10.1007/s10021-001-0077-1.

Peterson, G., Craig, R. A., & Holling, C. S. (1998). Ecological resilience, biodiversity, and scale. *Ecosystems, 1*, 6–18. https://doi.org/10.1007/s100219900002.

Pimm, S. L. (1991). The balance of nature? Ecological issues in the conservation of species and communities. Chicago: University of Chicago Press.

Reuter, H., Hölker, F., Middelhoff, U., Jopp, F., Eschenbach, C., & Breckling, B. (2005). The concepts of emergent and collective properties in individual-based models—Summary and outlook of the Bornhöved case studies. *Ecological Modelling, 186*, 489–501. https://doi.org/10.1016/j.ecolmodel.2005.02.014.

Rosenzweig, M. L. (1991). Habitat selection and population interactions: The search for mechanism. *The American Naturalist, 137*, S5–S28. https://doi.org/10.1086/285137.

Scalera, R. (2010). How much is Europe spending on invasive alien species? *Biological Invasions, 12*, 173–177. https://doi.org/10.1007/s10530-009-9440-5.

Smit, B., & Wandel, J. (2006). Adaptation, adaptive capacity and vulnerability. *Global Environmental Change, 16*, 282–292. https://doi.org/10.1016/j.gloenvcha.2006.03.008.

Steffen, W., Richardson, K., Rockström, J., Cornell, S. E., Fetzer, I., Bennett, E. M., et al. (2015). Planetary boundaries: Guiding human development on a changing planet. *Science, 347*, 1259855. https://doi.org/10.1126/science.1259855.

Sutherland, J. P. (1974). Multiple stable points in natural communities. *American Naturalist, 108*, 859–873.

Taylor, R. A., White, A., & Sherratt, J. A. (2013). How do variations in seasonality affect population cycles? *Proceedings of the Royal Society B: Biological Sciences, 280*, 20122714–20122714. https://doi.org/10.1098/rspb.2012.2714.

Thrush, S. F., Hewitt, J. E., Dayton, P. K., Coco, G., Lohrer, A. M., Norkko, A., et al. (2009). Forecasting the limits of resilience: Integrating empirical research with theory. *Proceedings of the Royal Society B: Biological Sciences, 276*, 3209–3217. https://doi.org/10.1098/rspb.2009.0661.

Turchin, P. (2003). Complex population dynamics. Princeton: Princeton University Press.

Turner, B. L., Matson, P. A., McCarthy, J. J., Corell, R. W., Christensen, L., Eckley, N., et al. (2003). Illustrating the coupled human–environment system for vulnerability analysis: Three case studies. *Proceedings of the National Academy of Sciences, 100*, 8080–8085. https://doi.org/10.1073/pnas.1231334100.

von Gleich, A., & Giese, B. (2019). Resilient systems as a biomimetic guiding concept. In M. Ruth & S. Gößling-Reisemann (Eds.), *Handbook on resilience of socio-technical systems* (p. 424). Edward Elgar Publishing.

von Gleich, A., Goessling-Reisemann, S., Stührmann, S., Woizeschke, P., & LutzKunisch, B. (2010). Resilienz als Leitkonzept-Vulnerabilität als analytische Kategorie. In K. Fichter, A. von Gleich, R. Pfriem, & B. Siebenhüner (Eds.), Theoretische Grundlagen für Klimaanpassungsstrategien, Nordwest2050-Berichte (pp. 13–49). Bremen/Oldenburg.

Walker, B. H. (1992). Biodiversity and ecological redundancy. *Conservation Biology, 6*, 7.

Walker, B. (1995). Conserving biological diversity through ecosystem resilience. *Conservation Biology, 9*, 747–752. https://doi.org/10.1046/j.1523-1739.1995.09040747.x.

Walker, B., Holling, C. S., Carpenter, S., & Kinzig, A. (2004). Resilience, adaptability and transformability in social–ecological systems. *Ecology and Society, 9*. https://doi.org/10.5751/ES-00650-090205.

Walter, J., Jentsch, A., Beierkuhnlein, C., & Kreyling, J. (2013). Ecological stress memory and cross stress tolerance in plants in the face of climate extremes. *Environmental and Experimental Botany, 94*, 3–8. https://doi.org/10.1016/j.envexpbot.2012.02.009.

Weißhuhn, P., Müller, F., & Wiggering, H. (2018). Ecosystem vulnerability review: Proposal of an interdisciplinary ecosystem assessment approach. *Environmental Management, 61*, 904–915. https://doi.org/10.1007/s00267-018-1023-8.

Werren, J. H. (2011). Selfish genetic elements, genetic conflict, and evolutionary innovation. *Proceedings of the National Academy of Sciences, 108*, 10863–10870. https://doi.org/10.1073/pnas.1102343108.

White, G. F. (1974). *Natural hazards*. New York: Oxford.

Whittaker, R. H., Levin, S., & Root, R. B. (1973). Niche, habitat, and ecotope. *American Naturalist, 107*, 321–338.

Williams, L. R. R., & Kapustka, L. A. (2000). Ecosystem vulnerability: A complex interface with technical components. *Environmental Toxicology and Chemistry, 19*, 1055–1058. https://doi.org/10.1002/etc.5620190435.

4

Bactrocera oleae: A Case Study

Merle Preu, Johannes L. Frieß, Broder Breckling and Winfried Schröder

Population Biology

The olive fruit fly *Bactrocera oleae* (Rossi) (Diptera, Tephritidae) (former name: *Dacus oleae*) is a phytophagous insect associated to olive trees (*Olea europaea*, Oleaceae). With its larvae feeding monophagously on olive fruits, the fly is considered the most severe pest of olive cultivation causing tremendous economic losses. The current distribution of *B. oleae* encompasses the Mediterranean basin, Africa, the Canary Islands, the Middle East, California and Central America (Daane and Johnson 2010; Nardi et al. 2005).

Phenology

Female olive fruit flies oviposit their eggs underneath the skin of ripening olive fruits (Nardi et al. 2005). Usually only one egg per fruit is laid, a process taking six to 13 min (Christenson and Foote 1960; Genç and Nation 2008a; Gutierrez et al. 2009). Studies found that females lay an average of four to 19 eggs per day and a total of 200–350 eggs in their lifetime (Genç and Nation 2008b; Kokkari et al. 2017; Sharaf 1980; Tsiropoulos 1980). To this end, undamaged and unripe olive fruits are preferred for oviposition (Genç and Nation 2008b; Yokoyama and Miller 2004). After hatching inside the fruit, larvae feed monophagously on the pulp and pass through three instar stages to complete larval development (Daane and Johnson 2010; Sharaf 1980). Prior

M. Preu · B. Breckling · W. Schröder
Landscape Ecology, University of Vechta, Vechta, Germany

J. L. Frieß (✉)
Institute of Safety/Security and Risk Sciences, University of Natural Resources and Life Sciences, Vienna (BOKU), Austria
e-mail: johannesfriess@gmx.net

to pupation, the last instar leaves the fruit, drops to the soil and pupates at a depth of one to nine centimetres (Dimou et al. 2003; Sharaf 1980). When conditions are favourable, e.g. during summer, pupation may also take place inside the olive fruit (Dimou et al. 2003; Sharaf 1980). Finally, mature adults emerge from the exuviae of the pupa.

Over the course of a year, the olive fruit fly exhibits three to five overlapping generations with seasonal fluctuations of fly densities (Boccaccio and Petacchi 2009; Comins and Fletcher 1988; Kokkari et al. 2017; Pontikakos et al. 2010; Voulgaris et al. 2013). The population dynamics of a Mediterranean population and phenological characteristics of investigated *B. oleae* populations are summarized in Table 4.1. In the Mediterranean, *B. oleae* eggs first appear in August, with a peak of oviposition occurring in mid-October (Bento et al. 1999; Kapatos and Fletcher 1984; Petacchi et al. 2015). Larvae are observed from the end of August and pupation was recorded from the middle of September onwards (Bento et al. 1999). However, life stage-specific developmental times and thus *B. oleae* population dynamics may considerably vary between locations and are strongly driven by climatic conditions (Marchi et al. 2016; Ordano et al. 2015).

While *B. oleae* may be able to even lay eggs in fall and early winter in mild environments (Castrignanò et al. 2012; Petacchi et al. 2015), harsh winter conditions induce overwintering for populations in most Mediterranean regions. The fly overwinters in its adult- or, more commonly, in its pupa-stage in the soil (Kapatos and Fletcher 1984; Neuenschwander et al. 1981; Sharaf 1980). Low winter temperatures regularly cause extremely high mortalities (Table 4.2), with reductions of the adult population of up to 99.7% (Arambourg and Pralavorio 1970; Bigler and Delucchi 1981; Gonçalves et al. 2012). A maximum of two flies per tree was determined to be the overwintering population in a Greek olive orchard (Krimbas and Tsakas 1971). Pupae disappear in high quantities of up to 98.5% over the winter (Arambourg and Pralavorio 1970; Gonçalves et al. 2012). Consequently, *B. oleae* is assumed to frequently go through a bottleneck during winter (Ochando and Reyes 2000). Since all populations in the same area suffer the bottleneck more or less simultaneously, a recolonization by surrounding populations is unlikely (Ochando and Reyes 2000). The spring population consists of the surviving adults, insects arising from overwintering pupae or individuals produced from a spring infestation of unharvested fruits or wild occurring olive trees (Economopoulos et al. 1982; Marchini et al. 2017).

Population Characteristics

In order to assess the inheritance of genetic features, certain qualities of the regarded population need to be considered. Therefore, this section reviews two important aspects determining gene propagation in *B. oleae* populations: sex ratio and mating behaviour.

Table 4.1 Summary of the population dynamics of Mediterranean *Bactrocera oleae* adult populations as reported in the scientific literature

Generations per year	Time of fly presence	Maximum fly density	Adult emerging from overwintering	Study area	References
3	July–late autumn			Tuscany, Italy	Boccaccio and Petacchi (2009)
4	June–November			Corfu, Greece	Kapatos and Fletcher (1984)
4			May	Tripolitania	Sharaf (1980)
			February/March	Arnasco and Levanto, Italy	Petacchi et al. (2015)
		April–May and August–October			Economopoulos et al. (1982)
		September–October		Spain	Ortega and Pascual (2014)
	April–November	October	April	Trasos-Montes, Portugal	Bento et al. (1999)

Table 4.2 Overwintering mortalities of different life stages of *Bactrocera oleae* reported in the literature

Life stage	Overwintering mortality (%)	References
Immature stages	98.5	Gonçalves et al. (2012)
	62.2–91.3	Bigler and Delucchi (1981)
Adults	98.5–99.7	Arambourg and Pralavorio (1970)

Sex Ratio

Trap catches constitute the major data source for population analyses of the olive fruit fly. The attractiveness of traps to olive fruit flies varies over the course of a year (Neuenschwander and Michelakis 1979a) with different effects of various trap systems (e.g. pheromone based traps, food odours) on the two sexes (Economopoulos and Stravropoulou-Delivoria 1984; Neuenschwander and Michelakis 1979a). These aspects need to be considered when looking at trap-determined sex ratios of wild populations. However, most studies report a sex ratio close to 1:1 (Ant et al. 2012; Moore 1962; Speranza et al. 2004). Katsoyannos and Kouloussis (2001) observed a slightly higher number of males (52.2% males) captured on coloured spheres over the course of two years and a total of 7518 flies. In contrast to these observations, in laboratory trials, adults emerged with a ratio of 1:2 (male:female) from reared larvae (Genç and Nation 2008b). Compared to trap catches, the knock-down method or so-called "sondage" (use of insecticides) is considered to be a relatively unbiased method for population analyses. With this technique, Neuenschwander and Michelakis (1979a) demonstrated that the sex ratio of wild populations stays closer to 1:1 than is suggested by trap catches.

Mating Behaviour

After emergence from the pupa, adults complete gonad maturation and reach sexual maturity within three to eight days (Canale et al. 2012; Mazomenos 1984). After mating, females remain unreceptive to further mating for a few days or weeks and store the received semen until ovulation (Solinas and Nuzzaci 1984; Tzanakakis et al. 1968; Zervas 1982). Studies demonstrated that female *B. oleae* are oligogamous, mating only one to three times (Zervas 1982; Zouros and Krimbas 1970). Zouros and Krimbas (1970) demonstrated monogamy for a vast proportion of female flies and estimated a frequency of 17% for female polygamy (Zouros and Krimbas 1970). In contrast to these findings, males are polygamous, mating daily when receptive females are available (Zervas 1982).

Strong intraspecific competition and sexual selection was observed for olive fruit flies (Benelli 2014; Benelli et al. 2012, 2013, 2015). Competition among male flies occurred for single leaf territories suitable for courtship display (wing vibration), whereas females competed for oviposition sites (Benelli 2014; Benelli et al. 2012). Aggressive behaviour was expressed by wing waving, fast running towards the opponent, pouncing and boxing on the head and thorax of the foe (Benelli 2014).

Environmental Tolerances

The life cycle of *Bactrocera oleae* is closely linked to environmental conditions, in particular to local climatic conditions (Fletcher et al. 1978). This section gives an overview on the two main climatic factors impacting the development of the olive fruit fly in Mediterranean regions: temperature and relative humidity.

Temperature

Local temperature is a major driver of olive fruit fly development and a determinant of the population dynamics (Marchi et al. 2016; Ordano et al. 2015). Consequently, many models on *B. oleae* phenology and population evolution are driven by local temperature data. For this, comprehensive knowledge on the effect of temperature on survival and development of different life stages of *B. oleae* is indispensable. A number of scientific studies have assessed temperature-dependent mortalities and developmental times of the olive fruit fly, e.g. in laboratory trials, the summarized data is displayed in Tables 4.3 and 4.4. In general, high mortalities of the fly are observed at high summer temperatures and at low temperatures over the winter months (Gonçalves et al. 2012). Thus, temperature conditions of the previous seasons were found to determine the rate of *B. oleae* infestation (Marchi et al. 2016). Mild winters are characteristically followed by high olive fruit fly infestations (Marchi et al. 2016).

Strong temperature effects have been observed on the reproduction of *Bactrocera oleae* (Tzanakakis and Koveos 1986). No ovarian development was observed at temperatures lower than 12 °C (Fletcher and Kapatos 1983). The upper maturation threshold was found to be 29.3 °C, with very few females maturing at 29 °C (Fletcher et al. 1978; Fletcher and Kapatos 1983). In some Mediterranean regions, these temperature requirements annually cause two periods when no reproduction occurs: during the winter months and at the beginning of summer, a period characterized by high temperatures and low humidity (Economopoulos et al. 1982; Fletcher et al. 1978). In mild climates, however, *B. oleae* might even be actively reproducing in winter (Economopoulos et al. 1982).

Relative Humidity

The development of *Bactrocera oleae*, i.e. longevity, maturation and survival, is strongly correlated with relative air humidity (Broufas et al. 2009; Broumas et al.

Table 4.3 Temperature-dependent mortalities of *Bactrocera oleae* life stages reported in the scientific literature

Life stage	Temperature (°C)	Mortality (proportion of population)	References
Egg	8	No hatching	Tsiropoulos (1972)
	10–30	<20%	Tsiropoulos (1972)
	16	45–51%	Genç and Nation (2008a)
	22	27–33%	Genç and Nation (2008a)
	25	13–31%	Sánchez-Ramos et al. (2013)
	27	13–26%	Genç and Nation (2008a)
	35	98%	Genç and Nation (2008a)
	35	No hatching	Tsiropoulos (1972)
Larvae	10	100%	Tsitsipis (1980)
	16	38–63%	Genç and Nation (2008a)
	22	18–47%	Genç and Nation (2008a)
	27	10–31%	Genç and Nation (2008a)
	32.5	100%	Tsitsipis (1980)
	35	92–94%	Genç and Nation (2008a)
Pupae	12.5	15.8%	Tsitsipis (1980)
	16	44–70%	Genç and Nation (2008a)
	22	26–56%	Genç and Nation (2008a)
	27	18–37%	Genç and Nation (2008a)
	30	48%	Tsitsipis (1980)
	35	No development	Genç and Nation (2008a)
Adult	16	53–76%	Genç and Nation (2008a)
	22	31–64%	Genç and Nation (2008a)
	27	25–50%	Genç and Nation (2008a)
	35	No development	Genç and Nation (2008a)

2002; Fletcher et al. 1978). Especially long life spans of adult flies have been observed in high air humidity (Broufas et al. 2009). A summary of the effect of humidity on different life parameters of *B. oleae* is displayed in Table 4.5. Humidity may also impact the effectiveness of specific trapping systems for fly monitoring (Bueno and Jones 2002; Mazomenos et al. 2002).

Dispersal Dynamics

Scientists have utilized various techniques to determine the dispersal potential of adult olive fruit flies, ranging from traditional capture–recapture experiments and

Table 4.4 Temperature-dependent development of *Bactrocera oleae* as reported in scientific literature

Stage	Temperature (°C)	Developmental time/Life span	References
Egg	10	3 days	Tsitsipis (1977)
	15	1 day	Tsitsipis (1977)
	16	11 days	Genç and Nation (2008a)
	20	1 day	Tsitsipis (1977)
	22	4 days	Genç and Nation (2008a)
	24	2 days	Genç and Nation (2008b)
	25	1 day	Tsitsipis (1977)
	27	3 days	Genç and Nation (2008a)
	30	1 day	Tsitsipis (1977)
	35	2 days	Genç and Nation (2008a)
Larvae	12.5	37 days	Tsitsipis (1980)
	15	25 days	Tsitsipis (1980)
	20	15 days	Tsitsipis (1980)
	16	33 days	Genç and Nation (2008a)
	22	14 days	Genç and Nation (2008a)
	25	10 days	Tsitsipis (1980)
	25	7 days	Neuenschwander and Michelakis (1979b)
	27	11 days	Genç and Nation (2008a)
	30	10 days	Tsitsipis (1980)
	35	5 days	Genç and Nation (2008a)
Pupa	12.5	49 days	Tsitsipis (1980)
	14	34.5 days	Tsiropoulos (1972)
	15	35 days	Tsitsipis (1980)
	16	28 days	Genç and Nation (2008a)
	20	15 days	Tsitsipis (1980)
	22	12 days	Genç and Nation (2008a)
	25	10 days	Tsitsipis (1980)
	25	13 days	Tsiropoulos (1972)
	27	8 days	Genç and Nation (2008a)
	30	9 days	Tsitsipis (1980)
	35	No development	Genç and Nation (2008a)
Adult (male)	25	47 days	Sánchez-Ramos et al. (2013)
Adult (female)	25	34 days	Sánchez-Ramos et al. (2013)
	25	19–31 days	Broufas et al. (2009)

Table 4.5 Life parameters of *Bactrocera oleae* depending on relative humidity as reported in scientific literature

Relative humidity	Temperature	Longevity	Proportion of mature females (%)	Mean number of eggs per female	Egg hatching (%)	References
12	25 °C	19 days	4	109	12	Broufas et al. (2009)
33		34 days	41	215	77	
55		48 days	61	444	82	
75		50 days	67	581	86	
94		31 days	92	234	87	
<90	20 °C				0	Tsitsipis and Abatzis (1980)
95					72	
100					97	

performance analyses on flight mills to the use of molecular techniques. This section provides a literature review in order to provide data appropriate for dispersal models. Experimentally determined maximum and typical dispersal distances as well as observed impacts of laboratory treatments and genetic parameters are summarized for Mediterranean *Bactrocera oleae* populations.

Dispersal Distances

In suitable environmental conditions, non-dispersive movements are assumed for adult olive fruit flies (Remund et al. 1976), with typical distances of 180–190 m overcome within two weeks (Fletcher and Economopoulos 1976; Fletcher and Kapatos 1981). Radioactive labelling revealed that the major proportion of a wild *B. oleae* population did not disperse further than 1000 m within one month and no migration was observed (Pelekassis et al. 1963). However, extreme conditions, such as high population densities (overcrowding) or lack of oviposition sites, might induce long distance dispersal of olive fruit flies (Economopoulos et al. 1978; Fletcher and Kapatos 1981; Remund et al. 1976). A release experiment by Fletcher and Kapatos (1981) illustrates this situation for *B. oleae* populations on Corfu, Greece. After release to a grove with no new season crop, flies travelled more than 400 m during the first seven days, but only dispersed over 180 m when 30% of olives in the release grove bore fruit (Fletcher and Kapatos 1981). Maximum dispersal distances for *B. oleae* reported in the literature range from 4000 to 5000 m (Economopoulos et al. 1978; Pelekassis et al. 1963; Remund et al. 1976). A summary of dispersal distances recorded in scientific studies is displayed in Table 4.6.

Table 4.6 Dispersal distances of *Bactrocera oleae* in the Mediterranean region

Distance (m)	Conditions	Duration	Survival (%)	References
50	Irradiated laboratory-reared olive fruit flies			Rempoulakis and Nestel (2012)
60	Laboratory-reared flies (olive diet)			Economopoulos et al. unpubl., reported in Remund et al. (1976)
180	30% fruit crop in release grove	1 week		Fletcher and Kapatos (1981)
180–190	Suitable environmental conditions, laboratory-reared flies	2 weeks	13	Fletcher and Economopoulos (1976)
400	No fruit crop in release grove	1 week		Fletcher and Kapatos (1981)
<1000	Wild flies, semi-mountainous olive grove, Greece, 86% of population dispersed up to 1000 m	35 days		Pelekassis et al. (1963)
1000–2500	Wild flies, semi-mountainous olive grove, Greece, 10% of population dispersed 1000–2500 m	35 days		Pelekassis et al. (1963)
2000	Wild flies, closest olive grove 2000 m distance		<1	Economopoulos et al. (1978)
2000	Artificially-reared flies, closest olive grove 2000 m distance		<1	Economopoulos et al. (1978)
2000–3000	Overcrowding, laboratory-reared flies (olive diet)			Economopoulos et al. unpubl., reported in Remund et al. (1976)
4000	Wild and artificially-reared flies, closest olive grove 2000 m distance		<1	Economopoulos et al. (1978)
2500–4300	Wild flies, semi-mountainous olive grove, Greece, 4% of population dispersed 2500–4300 m, maximum dispersal distance	35 days		Pelekassis et al. (1963)
3000–5000	Laboratory-reared flies (olive diet)			Economopoulos et al. unpubl., reported in Remund et al. (1976)

The flight potential of *B. oleae* is likely to vary with fly characteristics, e.g. fly age and sex. Remund et al. (1976) determined flight performance by means of flight propensity, number of flights and total distance flown for laboratory-reared *B. oleae* on flight-mills. The authors observed a threefold increase of the distance overcome by 14-day-old adult flies compared to their 2-day-old conspecifics. At the same time, females displayed a higher flight capacity with regard to all parameters assessed than their male counterparts. This effect of fly sex, however, was not detected in a release experiment by Fletcher and Economopoulos (1976).

Some strategies to reduce population densities of *B. oleae* in the Mediterranean area depend on the release of laboratory-reared individuals that have been sterilised by irradiation treatments. As a consequence, experimenters investigated the impact of laboratory-rearing as well as different rearing conditions on the fly's dispersal potential, assessing both non-sterilised and sterilised specimens. Based on results of field experiments, Fletcher and Economopoulos (1976) suggested greater dispersal distances for wild-compared to laboratory-reared olive fruit flies. In their experiment, no additional negative impact of the irradiation procedure of laboratory-reared *B. oleae* was observed. Remund et al. (1976) found a significant impact of diet on flight performance of olive fruit flies. Compared to individuals reared on an artificial diet, flies feeding on olives flew 2–3 times greater distances (Remund et al. 1976).

Genetic Variability and Gene Flow

Genetic parameters of *Bactrocera oleae* indicate a high degree of genetic variability for Mediterranean populations, as displayed by polymorphism and the number of alleles per locus (Augustinos et al. 2005; Ochando and Reyes 2000; Segura et al. 2008). Compared to polyphagous fruit flies, a lower genetic variability was expected for *B. oleae* due to its high host specialization. However, actual genetic variability was comparable or even higher (Augustinos et al. 2005; Ochando and Reyes 2000). At the same time, analyses detected more variability within than among Mediterranean *B. oleae* populations, indicating the occurrence of gene flow among these populations (Segura et al. 2008). In the regarded studies, gene flow (N*m*, with N as the effective population size and m as the net migration rate per generation) was estimated after Wright (1931). Despite the relatively short regular dispersal distances of *B. oleae* individuals observed in field experiments, molecular analyses indicate a high level of gene flow among Mediterranean populations (Augustinos et al. 2005; Ochando and Reyes 2000; Segura et al. 2008). Gene flow was as high as $Nm = 8.9$ between different populations from Spain that were separated by hundreds of kilometres of dry climate (Ochando and Reyes 2000). According to Wright (1931), an N*m* equal to 1 would be sufficient to prevent differentiation among populations. In addition to ecological dispersal, social-ecological aspects, such as trade with olive cultivars, have to be considered as potential source of gene flow (Augustinos et al. 2005; Segura et al. 2008).

With regard to the fly's dependence on its olive host and its winter disappearance, a high potential for the occurrence of bottlenecks was acknowledged in different studies (Augustinos et al. 2005; Ochando and Reyes 2000). As bottlenecks might result in inbreeding, heterozygote deficiency detected in Spanish fly populations substantiated this assumption (Ochando and Reyes 2000). An interesting result was an inhomogeneity of genetic variability among loci, which indicates selective processes (Augustinos et al. 2005). Natural selection due to agricultural practices may be the factor responsible for the pattern of genetic variability (Ochando and Reyes 2000).

Eggs

Hosted underneath the olive's epidermis, *B. oleae* eggs are protected from most natural enemies (Bateman 1972; Daane and Johnson 2010). However, predation may occur by birds feeding on whole olives, such as thrushes (Bigler et al. 1986; Neuenschwander et al. 1983), or by the midge *Prolasioptera berlesiana* (Kalaitzaki et al. 2014). Larvae of the latter occur in olive punctures infested with the fungus *Macrophoma dalmatica* and were observed to destroy olive fly eggs (Solinas 1967). Recognized as a so-called biocomplex, relationships and potential dependencies between *B. oleae*, *M. dalmatica* and *P. berlesiana* have extensively been studied and tested for the use in biological pest management. Even though olive fly oviposition holes appear to be the main entrance for the fungus as well as for the midge (González et al. 2006; Neuenschwander et al. 1983), neither of the insects plays an important role as disseminating agent of *M. dalmatica* spores (Harpaz and Gerson 1966). According to Neuenschwander et al. (1983) *P. berlesiana* oviposition immediately followed *B. oleae* attacks. The midges may, to a minor extent, prey on olive fly eggs and young larvae, but rely on fungal diet for survival (Harpaz and Gerson 1966; Solinas 1967). Maximum *B. oleae* egg mortality induced by *P. berlesiana* was estimated at 30–50% for Cretan populations (Neuenschwander et al. 1983). Overall predation of eggs, however, is low, with 0–6% predation-induced egg mortality determined for traditional groves in Portugal (Gonçalves et al. 2012).

Larvae

After hatching, *B. oleae* larvae feed monophagous on the pulp of their olive hosts (Daane and Johnson 2010). Within the tissue of an unripe fruit, olive fruit flies depend on the presence of symbiotic microorganisms for growth and survival (Bateman 1972; Ben-Yosef et al. 2010, 2014). The identity of the fly's symbiotic microflora is in the focus of scientific controversy. For many decades a mutualistic relationship with the extracellular bacterium *Pseudomonas savastanoi* has been suspected (Hagen 1966). However, most recently discovered associations involve the bacterium *Acetobacter tropicalis* (Kounatidis et al. 2009) and the before undescribed gut bacterium *Candidatus Erwinia dacicola,* which was numerically dominant inhabitant of the fly's oesophagus bulb (Ben-Yosef et al. 2014; Capuzzo et al. 2005; Estes et al. 2009;

Sacchetti et al. 2008). Based on the absence *of* Ca. *E. dacicola* in other investigated tephritids and the peculiar morphological structure of the bacteria-hosting oesophagus bulb of *B. oleae*, Capuzzo et al. (2005) proposed a coevolution of *B. oleae* with Ca.*E. dacicola.* A total of 16 different bacteria were found to be associated with wild olive flies in Greece, displaying varying bacteria compositions between generations of the same year (Tsiropoulos 1983). It is important to note, that a lower number of different bacteria, as well as a reduced compositional variability was detected in laboratory-reared olive flies (Tsiropoulos 1983).

Life Stages at the Soil

Life stages associated to the soil, such as late instar larvae, pupae and emerging adults, are most exposed to predation (Bateman 1972; Daane and Johnson 2010). Common predators at the soil surface are arthropods like beetles, myriapods and ants (Daane and Johnson 2010; Neuenschwander et al. 1983). Ants are considered to play a major role in the destruction of *B. oleae* developmental stages. Individual ants or groups are capable of transporting *B. oleae* to the nest, eventually damaging their prey (Neuenschwander et al. 1983). Ants were also reported to be the main predators of *B. oleae* pupae in recently invaded habitats in California (Orsini 2006; Orsini et al. 2007). Additionally, food searching birds, e.g. the black bird (*Turdus merula*), the robin (*Erithacus rubecula*) or the starling (*Sturnus vulgaris*), may destroy life stages of *B. oleae* on the soil surface (Neuenschwander et al. 1983). Pupae removal by birds can account for up to 70%, depending on orchard type (Bigler et al. 1986).

Adults

In contrast to their monophagous larvae, adults of *B. oleae* feed on a variety of organic sources, including insect honeydews, plant nectar and pollen, fruit exudates, bird dung, bacteria and yeast (Christenson and Foote 1960; Daane and Johnson 2010; Tsiropoulos 1977). As a result, adult olive flies may be found in vegetation other than olive trees, e.g. walnut, apple or plane trees (Economopoulos et al. 1982). Despite this range of food sources, the diet of *B. oleae* may be poor or unbalanced in its amino acid composition (Ben-Yosef et al. 2010). Experimental evidence suggests that the gut symbiontic bacterium Ca.*E. dacicola*, that was present in all life stages of *B. oleae* (Estes et al. 2009) enables adult olive fruit flies to utilize non-essential amino acids and urea as a source of nitrogen (Ben-Yosef et al. 2010, 2014). The authors proposed that *B. oleae* depends on the symbiont for protein synthesis and egg production.

Birds as well as cursorial and web spiders are considered to be important predators in olive orchards, capable of suppressing *B. oleae* population densities (Bigler et al. 1986; Picchi et al. 2016). These predators were found to be considerably more numerous around organic olive orchards compared to conventionally managed groves

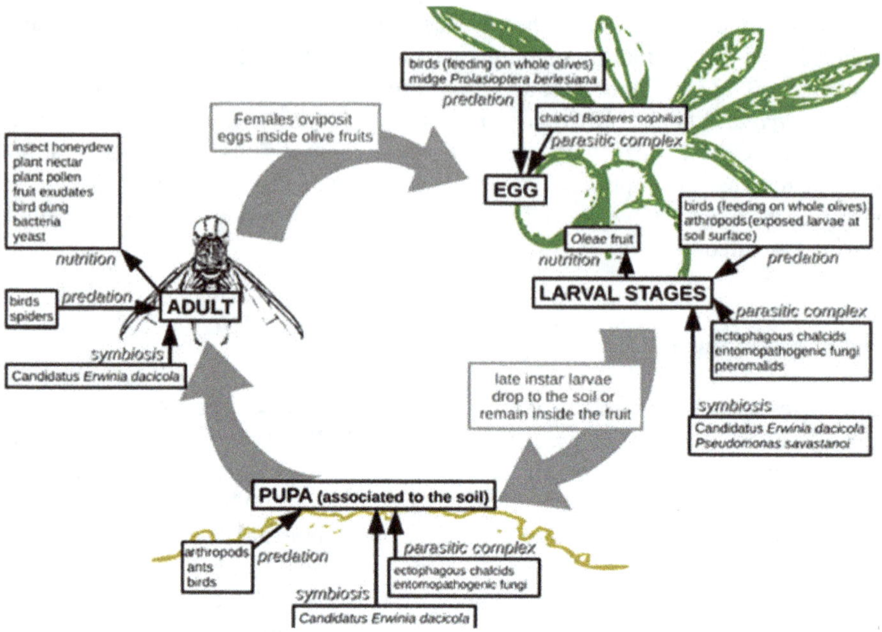

Fig. 4.1 Schematic representation of the ecological network of the olive fruit fly *Bactrocera oleae*

(Bigler et al. 1986; Picchi et al. 2016, 2017). However, the general lack of scientific literature on further potential predators of adult olive flies and predation rates highlights avenues for future research.

The life stages of *B. oleae* and interspecies relations are depicted in Fig. 4.1.

Competition for Olive Fruits

A number of olive tree pests have been identified besides the olive fruit fly. A comprehensive list of pests and diseases occurring on the Maltese Islands is provided by Haber and Mifsud (2007). The authors report different species of insects and eriophyid mites, as well as fungal and bacterial diseases. Most of these species occur primarily on leaves, stems and fresh shoots of the trees (Haber and Mifsud 2007; Martin et al. 2000), which indicates a rather low potential for direct competition for olive fruits with *B. oleae*. However, some hemiptera species also directly impact olive fruits, e.g. *Pollinia pollini*, *Aspidiotus nerii*, *Hemiberlesia rapax*, *Leucaspis riccae* (Alford 2014; Daane et al. 2005; Haber and Mifsud 2007). Even though these species do not penetrate the olive fruit skin, potential surface modifications might impede oviposition by *B. oleae*. In general, no literature references could be detected concerning the fly's competition with other olive pests.

At the same time, different studies highlight aggressive intraspecific interactions for adult olive fruit flies (Benelli 2014; Benelli et al. 2013, 2015). The intense intraspecific competition indicates that genetically modified flies, which might be released for eradication purposes, need to be capable of competing with wild types in order to mate and successfully spread the modified gene.

Potential Hazards with Regard to Gene Drive Release

Based on the ecological network analysis, a number of potential hazards, associated to the release of gene drive-equipped olive fruit flies to wild populations can be identified and are briefly discussed in this section.

Unintentional Long-Distance Transport of the Genetically Modified Organism

A serious increase of exposure associated with the release of genetically modified organisms is the potential for unintentional transport to non-target locations. Specimens might arrive in new locations via natural dispersal or by human agency. The latter bears potential for long-distance transport of the altered organism. In the receiving habitat, the newcomer may become invasive or co-exist with related species, enabling a potential transfer of modified genes across species boundaries. In this section potential pathways for long-distance transport of the olive fruit fly *Bactrocera oleae* are evaluated.

There are three main vectors for unintentional long-distance transport of *B. oleae*:

- Transport of adult individuals
  Living adult insect specimens might be transported as hitchhikers on aircrafts, e.g. in the passenger area or within the airline baggage. A prominent example of insect transport along this pathway is the commonly known airport malaria (Isaäcson 1989). In a similar manner, also agricultural pests may travel by this vector (Liebhold et al. 2006). The increase of public air traffic is expected to enhance the chances of insect transport by this vector in coming decades.
- Transport of immature life stages within the olive host
  The olive fruit fly undergoes egg and larval development within its olive fruit host. During this time, transport of the fruit might, in favourable conditions result in transport of the living insect to distant locations. Inspection records of 1962 indeed revealed that, within its host, the olive fruit fly arrived in the US via ships and aircrafts (Rainwater 1963). Both transport of single olives or entire trees may serve as a vector for unintentional introductions. Especially the trade of olive cultivars may result in unintentional introductions of the olive fruit fly and was suggested for the arrival of *B. oleae* in the New World (Nardi et al. 2005).

- Transport of immature stages (larvae, pupae) within the soil
  Life stages associated to the soil, such as larvae and pupae, may be transported within their host material. Soil adhering to shoes of outgoing tourists or soil associated to traded olive trees might serve as vector for immature stages of the olive fruit fly.

Hybridization and Horizontal Gene Transfer Across Species Boundaries

Ecological risk assessment for the release of gene drive-carrying organisms to natural populations is urged to consider the risk of unintentional transfer of the altered gene across species boundaries. Vertical inter-species transfer of altered genes will depend on the organism's ability to produce hybrids with related species.

Three fruit flies from the *B.* (subgenus *Bactrocera*) *dorsalis* complex (*B. dorsalis*, *B. papayae*, and *B. philippinensis*) were found capable of cross-mating (Schutze et al. 2013) and sperm transfer was observed between *B. dorsalis* and *B.* (*Bactrocera*) *carambolae* in outdoor field cages (McInnis et al. 1999). For the olive fruit fly, no studies testing the hybridization potential with related species were detected. *B. oleae* belongs to the subgenus *Daculus*, a sister group to the subgenus *Bactrocera* (Smith et al. 2003; Zhang et al. 2010). Hybridization trials should focus on the nearest African congeners (*Daculus*), *B. biguttula* and *B. munroi*, which are also associated with *Oleaceae* (Copeland et al. 2004; Drew and Hancock 2000) and exhibited a mean genetic divergence of 3.4% to *B. oleae* (Bon et al. 2016). Morphological features, i.e. length of terminalia, pheromone composition and variation in courtship signals might play a role for mating compatibility (Iwaizumi et al. 1997; Schutze et al. 2013).

The classification of Tephritidae species and their geographic distribution is shown in Fig. 4.2.

Ecological Niche Filling by Other Species

The declared aim of planned releases of gene drive-equipped insects is the local suppression or eradication of pest species. However, after extinction, the vacant ecological niche might be occupied by a related or competing species. In order to assess this possibility for the case study of the olive fruit fly, potential competitors of the olive fly were investigated. Despite the existence of other olive pests, no studies could be detected on competition with other olive pests. As a consequence of this potential lack of direct competitors, a genetically induced reduction in olive fruit fly densities may be unlikely to cause another pest population to increase.

Fig. 4.2 Classification and distribution of 17 Tephritidae species. Adapted from Bonizzoni et al. (2007)

The close co-evolution of *B. oleae* with symbiotic microorganisms, which is essential for its development inside an unripe fruit, already indicates that the ecological niche of the olive fruit fly might not easily be occupied by another organism.

Concluding Remark

With regard to its physiological and ecological characteristics, the olive fly can be considered a prototypic case to discuss the implications of SPAGE application in an economically significant, European context. Furthermore, *B. oleae* is a suitable candidate for a potential target organism in suppression drives, as it already has been the object of pest control efforts for several decades. Compared to other cases, the narrow specialisation and the involved network of ecological relations make this case relatively easy to survey. It appears well suitable to expand risk assessment experience with a wild species with a high dispersal potential, high rates of gene flow and even some risk of hybridisation with related species. It is easy to study under laboratory conditions and reproduces quickly. This case study revealed that uncertainties exist with regard to the dispersal capacity of gene drive-bearing olive flies, as well as concerning the high gene flow between different populations and most importantly with regard to the population bottlenecks that regularly occur in winter. These would significantly increase or decrease genetic variability between subpopulations and thereby severely jeopardize the intended outcome of any SPAGE-application. Nevertheless, there remain significant knowledge gaps in particular on the role of relations to symbiotic and other microflora and wider aspects of the ecological relations. However, experience gained in the study of this species can be useful to compare assessments of organisms with a more complex or more difficult to survey life cycles for which gene drive applications are discussed.

The Control of pest populations is one of the main objectives for possible SPAGE applications. The olive fly, as the main pest in olive cultivation poses a good example of a potential target organism in a European context. This case study revealed that uncertainties exist with regard to the dispersal capacity of gene drive-bearing olive flies, as well as concerning the high gene flow between different populations and most importantly with regard to the population bottlenecks that regularly occur in winter. These would significantly increase or decrease genetic variability between subpopulations and thereby severely jeopardize the intended outcome of any SPAGE-application. In summary, it can be said that the naturally occurring variability, in contrast to comparatively homogeneous laboratory conditions, leads to considerable and easily underestimated uncertainties about possible post-release effects. This is further exemplified in the stochastic model concerning winter bottlenecks and the individual based model, explored in Chap. 6. These models clearly show that even with a highly efficient gene drive, fluctuations in a single ecological factor may lead to vastly different outcomes.

Acknowledgements We would like to thank all the stakeholders for their continued interest in this project. We are especially grateful to Dr. Emmanouil Kabourakis from the Institute of viticulture, floriculture and vegetable crops (IVFVC), National Agricultural Research Foundadation (NAGREF) for his input to the project and particularly this case study.

References

Alford, D. V. (2014). *Pests of fruit crops: A colour handbook* (2nd Ed.). Boca Raton: CRC Press.

Ant, T., Koukidou, M., Rempoulakis, P., Gong, H.-F., Economopoulos, A., Vontas, J., & Alphey, L. (2012). Control of the olive fruit fly using genetics-enhanced sterile insect technique. *BMC Biology, 10*, 51. https://doi.org/10.1186/1741-7007-10-51.

Arambourg, Y., & Pralavorio, R. (1970). Survival of *Dacus oleae* Gmel. during the winter. *Annales de Zoologie Ecologie Animale, 2*, 659–622.

Augustinos, A. A., Mamuris, Z., Stratikopoulos, E., D'Amelio, S., Zacharopoulou, A., & Mathiopoulos, K. D. (2005). Microsatellite analysis of olive fly populations in the Mediterranean indicates a westward expansion of the species. *Genetica, 125*, 231–241. https://doi.org/10.1007/s10709-005-8692-y.

Bateman, M. A. (1972). The ecology of fruit flies. *Annual Review of Entomology, 17*, 493–518. https://doi.org/10.1146/annurev.en.17.010172.002425.

Benelli, G. (2014). Aggressive behavior and territoriality in the olive fruit fly, *Bactrocera oleae* (Rossi) (Diptera: Tephritidae): Role of residence and time of day. *Journal of Insect Behavior, 27*, 145–161.

Benelli, G., Bonsignori, G., Stefanini, C., Raspi, A., & Canale, A. (2013). The production of female sex pheromone in *Bactrocera oleae* (Rossi) young males does not influence their mating chances. *Entomological Science, 16*, 47–53. https://doi.org/10.1111/j.1479-8298.2012.00538.x.

Benelli, G., Canale, A., Bonsignori, G., Ragni, G., Stefanini, C., & Raspi, A. (2012). Male wing vibration in the mating behavior of the olive fruit fly *Bactrocera oleae* (Rossi) (Diptera: Tephritidae). *Journal of Insect Behavior, 25*, 590–603.

Benelli, G., Desneux, N., Romano, D., Conte, G., Messing, R. H., & Canale, A. (2015). Contest experience enhances aggressive behaviour in a fly: When losers learn to win. *Scientific Reports, 5*, 9347. https://doi.org/10.1038/srep09347.

Bento, A., Torres, L., Lopes, J., & Sismeiro, R. (1999). A contribution to the knowledge of *Bactrocera oleae* (GMEL) in Tras-os-Montes region (Northeastern Portugal): Phenology, losses and control. *Acta Horticulture, 474*, 541–544.

Ben-Yosef, M., Aharon, Y., Jurkevitch, E., & Yuval, B. (2010). Give us the tools and we will do the job: Symbiotic bacteria affect olive fly fitness in a diet-dependent fashion. *Proceedings of the Royal Society B: Biological Sciences, 277*, 1545–1552. https://doi.org/10.1098/rspb.2009.2102.

Ben-Yosef, M., Pasternak, Z., Jurkevitch, E., & Yuval, B. (2014). Symbiotic bacteria enable olive flies (*Bactrocera oleae*) to exploit intractable sources of nitrogen. *Journal of Evolutionary Biology, 27*, 2695–2705. https://doi.org/10.1111/jeb.12527.

Bigler, F., & Delucchi, V. (1981). Evaluation of the prepupal mortality of the olive fly, *Dacus oleae* Gmel. (Dipt., Tephritidae), on oleasters and olive trees in western Crete Greece. *Journal of Applied Entomology, 92*, 189–201.

Bigler, F., Neuenschwander, P., Delucchi, V., & Michelakis, S. (1986). Natural enemies of preimaginal stages of *Dacus oleae* Gmel. (Dipt., Tephritidae) in Western Crete. II. Impact on olive fly populations. *Bollettino del Laboratorio di Entomologia Agraria Filippo Silvestri, Portici, 43*, 79–96.

Boccaccio, L., & Petacchi, R. (2009). Landscape effects on the complex of *Bactrocera oleae* parasitoids and implications for conservation biological control. *BioControl, 54*, 607–616. https://doi.org/10.1007/s10526-009-9214-0.

Bon, M. C., Hoelmer, K. A., Pickett, C. H., Kirk, A. A., He, Y., Mahmood, R., & Daane, K. M. (2016). Populations of *Bactrocera oleae* (Diptera: Tephritidae) and its parasitoids in Himalayan Asia. *Annals of the Entomological Society of America, 109*, 81–91. https://doi.org/10.1093/aesa/sav114.

Bonizzoni, M., Gomulski, L. M., Malacrida, A. R., Capy, P., & Gasperi, G. (2007). Highly similar piggyBac transposase-like sequences from various Bactrocera (Diptera, Tephritidae) species. *Insect Molecular Biology, 16*, 645–650. https://doi.org/10.1111/j.1365-2583.2007.00756.x.

Broufas, G. D., Pappas, M. L., & Koveos, D. S. (2009). Effect of relative humidity on longevity, ovarian maturation, and egg production in the olive fruit fly (Diptera: Tephritidae). *Annals of the Entomological Society of America, 102*, 70–75. https://doi.org/10.1603/008.102.0107.

Broumas, T., Haniotakis, G., Liaropoulos, C., Tomazou, T., & Ragoussis, N. (2002). The efficacy of an improved form of the mass-trapping method, for the control of the olive fruit fly, *Bactrocera oleae* (Gmelin) (Dipt., Tephritidae): Pilot-scale feasibility studies. *Journal of Applied Entomology, 126*, 217–223. https://doi.org/10.1046/j.1439-0418.2002.00637.x.

Bueno, A. M., & Jones, O. T. (2002). Alternative methods for controlling the olive fly *Bactrocera olea* involving semiochemicals. *IOBC/WPRS Bulletin, 25*, 1–11.

Canale, A., Carpita, A., Conti, B., Canovai, R., & Raspi, A. (2012). Effect of age on 1,7-dioxaspiro-[5.5]-undecane production in both sexes of olive fruit fly, *Bactrocera oleae* (Diptera Tephritidae). *IOBC/WPRS Bulletin*, 219–225.

Capuzzo, C., Firrao, G., Mazzon, L., Squartini, A., & Girolami, V. (2005). "*Candidatus Erwinia dacicola*", a coevolved symbiotic bacterium of the olive fly *Bactrocera oleae* (Gmelin). *International Journal of Systematic and Evolutionary Microbiology, 55*, 1641–1647. https://doi.org/10.1099/ijs.0.63653-0.

Castrignanò, A., Boccaccio, L., Cohen, Y., Nestel, D., Kounatidis, I., Papadopoulos, N.T., et al. (2012). Spatio-temporal population dynamics and area-wide delineation of *Bactrocera oleae* monitoring zones using multi-variate geostatistics. *Precision Agriculture, 13*, 421–441. https://doi.org/10.1007/s11119-012-9259-4.

Christenson, L. D., & Foote, R. H. (1960). Biology of fruit flies. *Annual Review of Entomology, 5*, 171–192.

Comins, H. N., & Fletcher, B. S. (1988). Simulation of fruit fly population dynamics, with particular reference to the olive fruit fly. *Ecological Modelling, 40*, 213–231.

Copeland, R. S., White, I. M., Okumu, M., Machera, P., & Wharton, R. A. (2004). Insects associated with fruits of the Oleaceae (Asteridae, Lamiales) in Kenya, with special reference to the Tephritidae (Diptera). *Bishop Museum Bulletin in Entomology, 12*, 135–164.

Daane, K. M., & Johnson, M. W. (2010). Olive fruit fly: Managing an ancient pest in modern times. *Annual Review of Entomology, 55*, 151–169. https://doi.org/10.1146/annurev.ento.54.110807.090553.

Daane, K. M., Rice, R. E., Zalom, F. G., Barnett, W. W., & Johnson, M. W. (2005). Arthropod pests of olive. In G. S. Sibbett & L. Ferguson (Eds.), *Olive production manual* (pp. 105–114). University of California, Agriculture and Natural Resources.

Dimou, I., Koutsikopoulos, C., Economopoulos, A. P., & Lykakis, J. (2003). Depth of pupation of the wild olive fruit fly, *Bactrocera* (*Dacus*) *oleae* (Gmel.) (Dipt., Tephritidae), as affected by soil abiotic factors. *Journal of Applied Entomology, 127*, 12–17. https://doi.org/10.1046/j.1439-0418.2003.00686.x.

Drew, R. A., & Hancock, D. L. (2000). Phylogeny of the tribe Dacini (Dacinae) based on morphological, distributional, and biological data. In M. Aluja & A. L. Norrbom (Eds.), *Fruit flies (Tephritidae): Phylogeny and evolution of behaviour* (pp. 491–504). Boca Raton: CRC Press.

Economopoulos, A., & Stravropoulou-Delivoria, A. (1984). Yellow sticky rectangle with ammonium acetate slow-release dispenser: An efficient long-lasting trap for *Dacus oleae*. *Entomologia Hellenica, 2*, 17–23.

Economopoulos, A. P., Haniotakis, G. E., Mathioudis, J., & Missis, N. (1978). Long-distance flight of wild and artificially-reared *Dacus oleae* (Gmelin) (Diptera, Tephritidae). *Journal of Applied Entomology, 87*, 101–108.

Economopoulos, A. P., Haniotakis, G. E., Michelakis, S., Tsiropoulos, G. J., Zervas, G. A., Tsitsipis, J. A., et al. (1982). Population studies on the olive fruit fly, *Dacus oleae* (Gmel.) (Dipt., Tephritidae) in Western Crete. *Journal of Applied Entomology, 93*, 463–476. https://doi.org/10.1111/j.1439-0418.1982.tb03621.x.

Estes, A. M., Hearn, D. J., Bronstein, J. L., & Pierson, E. A. (2009). The olive fly endosymbiont, "*Candidatus Erwinia dacicola*", switches from an intracellular existence to an extracellular existence during host insect development. *Applied and Environment Microbiology, 75*, 7097–7106. https://doi.org/10.1128/AEM.00778-09.

Fletcher, B. S., & Economopoulos, A. P. (1976). Dispersal of normal and irradiated laboratory strains and wild strains of the olive fly *Dacus oleae* in an olive grove. *Entomologia Experimentalis et Applicata, 20*, 183–194.

Fletcher, B. S., & Kapatos, E. (1981). Dispersal of the olive fly, *Dacus oleae*, during the summer period on Corfu. *Entomologia Experimentalis et Applicata, 29*, 1–8.

Fletcher, B. S., & Kapatos, E. T. (1983). The influence of temperature, diet and olive fruits on the maturation rates of female olive flies at different times of the year. *Entomologia Experimentalis et Applicata, 33*, 244–252. https://doi.org/10.1111/j.1570-7458.1983.tb03264.x.

Fletcher, B. S., Pappas, S., & Kapatos, E. (1978). Changes in the ovaries of olive flies (*Dacus oleae* (Gmelin)) during the summer, and their relationship to temperature, humidity and fruit availability. *Ecological Entomology, 3*, 99–107. https://doi.org/10.1111/j.1365-2311.1978.tb00908.x.

Genç, H., & Nation, J. L. (2008a). Survival and development of Bactrocera oleae Gmelin (Diptera: Tephritidae) immature stages at four temperatures in the laboratory. *African Journal of Biotechnology, 7*, 2495–2500.

Genç, H., & Nation, J. L. (2008b). Maintaining *Bactrocera oleae* (Gmelin.) (Diptera: Tephritidae) colony on its natural host in the laboratory. *Journal of Pest Science, 81*, 167–174. https://doi.org/10.1007/s10340-008-0203-3.

Gonçalves, F. M., Rodrigues, M. C., Pereira, J. A., Thistlewood, H., & Torres, L. M. (2012). Natural mortality of immature stages of *Bactrocera oleae* (Diptera: Tephritidae) in traditional olive groves from north-eastern Portugal. *Biocontrol Science and Technology, 22*, 837–854. https://doi.org/10.1080/09583157.2012.691959.

González, N., Trapero, A., & Vargas Osuna, E. (2006). Dalmatian disease of olive fruits, 1: Biology and damages in olive orchards of the Seville province [Spain]. *Boletin de Sanidad Vegetal Plagas*.

Gutierrez, A. P., Ponti, L., & Cossu, Q. A. (2009). Effects of climate warming on Olive and olive fly (*Bactrocera oleae* (Gmelin)) in California and Italy. *Climatic Change, 95*, 195–217. https://doi.org/10.1007/s10584-008-9528-4.

Haber, G., & Mifsud, D. (2007). Pests and diseases associated with olive trees in the Maltese Islands (Central Mediterranean). *The Central Mediterranean Naturalist, 4*, 143–161.

Hagen, K. S. (1966). Dependence of the olive fly, *Dacus oleae*, larvae on symbiosis with *Pseudomonas savastanoi* for the utilization of olive. *Nature, 209*, 143–146.

Harpaz, I., & Gerson, U. (1966). The "biocomplex" of the olive fruit fly (*Dacus oleae* Gmel.), the olive fruit midge (*Prolasioptera berlesiana* Paoli), and the fungus *Macrophoma dalmática* Berl. & Vogl. in olive fruits in the Mediterranean basin (pp. 81–126, pp. ref.31/2 pp).

Isaäcson, M. (1989). Airport malaria: A review. *Bulletin of the World Health Organization, 67*, 737–743.

Iwaizumi, R., Kaneda, M., & Iwahashi, O. (1997). Correlation of length of terminalia of males and females among nine species of *Bactrocera* (Diptera: Tephritidae) and differences among sympatric species of *B. dorsalis* complex. *Annals of the Entomological Society of America, 90*, 664–666.

Kalaitzaki, A., Perdikis, D., Marketaki, M., Gyftopoulos, N., & Paraskevopoulos, A. (2014). Natural enemy complex of *Bactrocera oleae* in organic and conventional olive groves. *IOBC/WPRS Bulletin, 108*, 61–68.

Kapatos, E. T., & Fletcher, B. S. (1984). The Phenology of the olive fly, *Dacus oleae* (Gmel.) (Diptera, Tephritidae), in Corfu. *Journal of Applied Entomology, 97*, 360–370. https://doi.org/10.1111/j.1439-0418.1984.tb03760.x.

Katsoyannos, B. I., & Kouloussis, N. A. (2001). Captures of the olive fruit fly *Bactrocera oleae* on spheres of different colours. *Entomologia Experimentalis et Applicata, 100*, 165–172. https://doi.org/10.1023/A:1019232623830.

Kokkari, A. I., Pliakou, O. D., Floros, G. D., Kouloussis, N. A., & Koveos, D. S. (2017). Effect of fruit volatiles and light intensity on the reproduction of *Bactrocera* (*Dacus*) *oleae*. *Journal of Applied Entomology*, 1–9. https://doi.org/10.1111/jen.12389.

Kounatidis, I., Crotti, E., Sapountzis, P., Sacchi, L., Rizzi, A., Chouaia, B., et al. (2009). *Acetobacter tropicalis* is a major symbiont of the olive fruit fly (*Bactrocera oleae*). *Applied and Environmental Microbiology, 75*, 3281–3288. https://doi.org/10.1128/AEM.02933-08.

Krimbas, C. B., & Tsakas, S. (1971). The genetics of *Dacus oleae*. V. Changes of esterase polymorphism in a natural population following insecticide control—selection or drift? *Evolution (N. Y), 25*, 454–460.

Liebhold, A. M., Work, T. T., McCullough, D. G., & Cavey, J. F. (2006). Airline baggage as a pathway for alien insect species invading the United States. *American Entomologist, 53*, 48–54. https://doi.org/10.1093/ae/52.1.48.

Marchi, S., Guidotti, D., Ricciolini, M., & Petacchi, R. (2016). Towards understanding temporal and spatial dynamics of *Bactrocera oleae* (Rossi) infestations using decade-long agrometeorological time series. *International Journal of Biometeorology, 60*, 1681–1694.

Marchini, D., Petacchi, R., & Marchi, S. (2017). *Bactrocera oleae* reproductive biology: New evidence on wintering wild populations in olive groves of Tuscany (Italy). *Bulletin of Insectology, 70*, 121–128.

Martin, J. H., Mifsud, D., & Rapisarda, C. (2000). The whiteflies (Hemiptera: Aleyrodidae) of Europe and the Mediterranean basin. *Bulletin of Entomological Research, 90*, 407–448.

Mazomenos, B. E. (1984). Effect of age and mating on pheromone production in the female olive fruit-fly, *Dacus oleae* (Gmel). *Journal of Insect Physiology, 30*, 765–769.

Mazomenos, B. E., Pantazi-Mazomenou, A., & Stefanou, D. (2002). Attract and kill of the olive fruit fly *Bactrocera oleae* in Greece as a part of an integrated control system. *IOBC/WPRS Bulletin, 25.*

McInnis, D. O., Rendon, P., Jang, E. B., van Sauers-Muller, A., Sugayama, R., & Malavasi, A. (1999). Interspecific mating of introduced, sterile *Bactrocera dorsalis* with wild *B. carambolae* (Diptera: Tephritidae) in Suriname: A potential case for cross-species sterile insect technique. *Annals of the Entomological Society of America, 92*, 758–765.

Moore, I. (1962). Further investigations on the artificial breeding of the olive fly—*Dacus oleae* GMEL.—under aseptic conditions (1). *Entomophaga, 7*, 53–57.

Nardi, F., Carapelli, A., Dallai, R., Roderick, G. K., & Frati, F. (2005). Population structure and colonization history of the olive fly, *Bactrocera oleae* (Diptera, Tephritidae). *Molecular Ecology, 14*, 2729–2738. https://doi.org/10.1111/j.1365-294X.2005.02610.x.

Neuenschwander, P., Bigler, F., Delucchi, V., & Michelakis, S. (1983). Natural enemies of preimmaginal stages of *Dacus oleae* Gmel. (Dipt., Tephritidae) in Western Crete. I. Bionomics and phenologies. *Bollettino del Laboratorio di Entomologia Agraria Filippo Silvestri, Portici, 40*, 3–32.

Neuenschwander, P., & Michelakis, S. (1979a). McPhail trap captures of *Dacus oleae* (Gmel.) (Diptera, Tephritidae) in comparison to the fly density and population composition as assessed by sondage technique in Crete, Greece. *Mitteilungen der Schweizerischen Entomologischen Gesellschaft, 52*, 343–357.

Neuenschwander, P., & Michelakis, S. (1979b). Determination of the lower thermal thresholds and day-degree requirements for eggs and larvae of *Dacus oleae* (Gmel.) (Diptera: Tephritidae) under field conditions in Crete, Greece. *Mitteilungen der Schweizerischen Entomologischen Gesellschaft.*

Neuenschwander, P., Michelakis, S., & Bigler, F. (1981). Abiotic factors affecting mortality of *Dacus oleae* larvae and pupae in the soil. *Entomologia Experimentalis et Applicata, 30*, 1–9. https://doi.org/10.1111/j.1570-7458.1981.tb03577.x.

Ochando, M. D., & Reyes, A. (2000). Genetic population structure in olive fly *Bactrocera oleae* (Gmelin): Gene flow and patterns of geographic differentiation. *Journal of Applied Entomology, 124*, 177–183. https://doi.org/10.1046/j.1439-0418.2000.00460.x.

Ordano, M., Engelhard, I., Rempoulakis, P., Nemny-Lavy, E., Blum, M., Yasin, S., et al. (2015). Olive fruit fly (*Bactrocera oleae*) population dynamics in the Eastern Mediterranean: Influence of exogenous uncertainty on a monophagous frugivorous Insect. *PLoS One, 10*, 1–18. https://doi.org/10.1371/journal.pone.0127798.

Orsini, M. (2006). Mortality and predation of olive fly (*Bactrocera oleae*) pupae on the soil in a Davis, California Olive Orchard (pp. 1–18).

Orsini, M., Daane, K. M., Sime, K. R., & Nelson, E. H. (2007). Mortality of olive fruit fly pupae in California. *Biocontrol Science and Technology, 17*, 797–807.

Ortega, M., & Pascual, S. (2014). Spatio-temporal analysis of the relationship between landscape structure and the olive fruit fly *Bactrocera oleae* (Diptera: Tephritidae). *Agricultural and Forest Entomology, 16*, 14–23. https://doi.org/10.1111/afe.12030.

Pelekassis, C. E. D., Mourikos, P. A., & Bantzios, D. N. (1963). Preliminary studies of the field movement of the olive fruit fly (*Dacus oleae* Gmel.) by labelling a natural population with radioactive phosphorus (P32). In *Radiation and radioisotopes applied to insects of agricultural importance* (pp. 105–114).

Petacchi, R., Marchi, S., Federici, S., & Ragaglini, G. (2015). Large-scale simulation of temperature-dependent phenology in wintering populations of *Bactrocera oleae* (Rossi). *Journal of Applied Entomology, 139*, 496–509. https://doi.org/10.1111/jen.12189.

Picchi, M. S., Bocci, G., Petacchi, R., & Entling, M. H. (2016). Effects of local and landscape factors on spiders and olive fruit flies. *Agriculture, Ecosystems & Environment, 222*, 138–147. https://doi.org/10.1016/j.agee.2016.01.045.

Picchi, M. S., Marchi, S., Albertini, A., & Petacchi, R. (2017). Organic management of olive orchards increases the predation rate of overwintering pupae of *Bactrocera oleae* (Diptera: Tephritidae). *Biological Control, 108*, 9–15.

Pontikakos, C. M., Tsiligiridis, T. A., & Drougka, M. E. (2010). Location-aware system for olive fruit fly spray control. *Computers and Electronics in Agriculture, 70*, 355–368. https://doi.org/10.1016/j.compag.2009.07.013.

Rainwater, H. I. (1963). Agricultural insect pest hitchhikers on aircraft. *Proceedings of the Hawaiian Entomological Society, 18*, 303–310.

Rempoulakis, P., & Nestel, D. (2012). Dispersal ability of marked, irradiated olive fruit flies [*Bactrocera oleae* (Rossi) (Diptera: Tephritidae)] in arid regions. *Journal of Applied Entomology, 136*, 171–180. https://doi.org/10.1111/j.1439-0418.2011.01623.x.

Remund, U., Boller, E. F., Economopoulos, A. P., & Tsitsipis, J. A. (1976). Flight performance of *Dacus oleae* reared on olives and artificial diet. *Journal of Applied Entomology, 82*, 330–339. https://doi.org/10.1111/j.1439-0418.1976.tb03420.x.

Sacchetti, P., Granchietti, A., Landini, S., Viti, C., Giovannetti, L., & Belcari, A. (2008). Relationships between the olive fly and bacteria. *Journal of Applied Entomology, 132*, 682–689. https://doi.org/10.1111/j.1439-0418.2008.01334.x.

Sánchez-Ramos, I., Fernández, C. E., González-Núñez, M., & Pascual, S. (2013). Laboratory tests of insect growth regulators as bait sprays for the control of the olive fruit fly, *Bactrocera oleae* (Diptera: Tephritidae). *Pest Management Science, 69*, 520–526. https://doi.org/10.1002/ps.3403.

Schutze, M. K., Jessup, A., Ul-Haq, I., Vreysen, M., Wornoayporn, V., Vera, M., et al. (2013). Mating compatibility among four pest members of the *Bactrocera dorsalis* fruit fly species complex (Diptera: Tephritidae). *Journal of Economic Entomology, 106*, 695–707. https://doi.org/10.1603/EC12409.

Segura, M. D., Callejas, C., & Ochando, M. D. (2008). *Bactrocera oleae*: A single large population in Northern Mediterranean basin. *Journal of Applied Entomology, 132*, 706–713. https://doi.org/10.1111/j.1439-0418.2008.01366.x.

Sharaf, N. S. (1980). Life history of the olive fruit fly, *Dacus oleae* Gmel. (Diptera: Tephritidae), and its damage to olive fruits in Tripolitania. *Journal of Applied Entomology, 89*, 390–400.

Smith, P. T., Kambhampati, S., & Armstrong, K. A. (2003). Phylogenetic relationships among *Bactrocera* species (Diptera: Tephritidae) inferred from mitochondrial DNA sequences. *Molecular Phylogenetics and Evolution, 26*, 8–17.

Solinas, M. (1967). Osservazioni biologiche condotte in puglia sulla Prolasioptera berlesiana Paoli, con particolare rife-rimento al rapporti simbiotici col *Dacus oleae* Gmel. e con la Sphaeropsis dalmatica (Thüm.) gigante. *Entomologica, 3*, 129–176.

Solinas, M., & Nuzzaci, G. (1984). Functional anatomy of *Dacus oleae* Gmel. female genitalia in relation to insemination and fertilization process. *Entomologica, 19*, 135–165. https://doi.org/10.15162/0425-1016/584.

Speranza, S., Bellocchi, G., & Pucci, C. (2004). IPM trials on attract-and-kill mixtures against the olive fly *Bactrocera oleae* (Diptera Tephritidae). *Bulletin of Insectology, 57*, 111–115.

Tsiropoulos, G. J. (1972). Storage temperatures for eggs and pupae of the olive fruit fly. *Journal of Economic Entomology, 65*, 100–102. https://doi.org/10.1093/jee/65.1.100.

Tsiropoulos, G. J. (1977). Reproduction and survival of the adult *Dacus oleae* feeding on pollens and honeydews. *Environmental Entomology, 6*, 390–392.

Tsiropoulos, G. J. (1980). Major nutritional requirements of adult *Dacus oleae*. *Annals of the Entomological Society of America, 73*, 251–253. https://doi.org/10.1093/aesa/73.3.251.

Tsiropoulos, G. J. (1983). Microflora associated with wild and laboratory reared adult olive fruit flies, *Dacus oleae* (Gmel.). *Journal of Applied Entomology, 96*, 337–340. https://doi.org/10.1111/j.1439-0418.1983.tb03680.x.

Tsitsipis, J. A. (1977). Effect of constant temperatures on eggs of olive fruit fly, *Dacus oleae* (Diptera: Tephritidae). *Annales de Zoologie Ecologie Animale, 9*, 133–139.

Tsitsipis, J. A. (1980). Effect of constant temperatures on larval and pupal development of olive fruit flies reared on artificial diet. *Environmental Entomology, 9*, 764–768. https://doi.org/10.1093/ee/9.6.764.

Tsitsipis, J. A., & Abatzis, C. (1980). Relative humidity effects, at 20 °C, on eggs of the olive fruit fly, *Dacus oleae* (Diptera: Tephritidae), reared on artificial diet. *Entomologia Experimentalis et Applicata, 28*, 92–99.

Tzanakakis, M. E., & Koveos, D. S. (1986). Inhibition of ovarian maturation in the olive fruit fly, *Dacus oleae* (Diptera: Tephritidae), under long photophase and an increase of temperature. *Annals of the Entomological Society of America, 79*, 15–18.

Tzanakakis, M. E., Tsitsipis, J. A., & Economopoulos, A. P. (1968). Frequency of mating in females of the olive fruit fly under laboratory conditions. *Journal of Economic Entomology, 61*, 1309–1312. https://doi.org/10.1093/jee/61.5.1309.

Voulgaris, S., Stefanidakis, M., Floros, A., & Avlonitis, M. (2013). Stochastic modeling and simulation of olive fruit fly outbreaks. *Procedia Technology, 8*, 580–586. https://doi.org/10.1016/j.protcy.2013.11.083.

Wright, S. (1931). Evolution in Mendelian populations. *Genetics*, 97–159. https://doi.org/10.1007/BF02459575.

Yokoyama, V. Y., & Miller, G. T. (2004). Quarantine strategies for olive fruit fly (Diptera: Tephritidae): Low-temperature storage, brine, and host relations. *Journal of Economic Entomology, 97*, 1249–1253. https://doi.org/10.1603/0022-0493-97.4.1249.

Zervas, G. (1982). Reproductive physiology of *Dacus oleae* (GMELIN) Diptera Tephritidae. Comparison of wild and laboratory reared flies. In *Geoponika*.

Zhang, B., Liu, Y. Y. H., Wu, W. W. X., & Wang, Z. Z. L. (2010). Molecular phylogeny of Bactrocera species (Diptera: Tephritidae: Dacini) inferred from mitochondrial sequences of 16S rDNA and COI sequences. *Florida Entomology, 93*, 369–377. https://doi.org/10.1653/024.093.0308.

Zouros, E., & Krimbas, C. B. (1970). Frequency of female digamy in a natural population of the olive fruit fly *Dacus oleae* as found using enzyme polymorphism. *Entomologia Experimentalis et Applicata, 13*, 1–9. https://doi.org/10.1111/j.1570-7458.1970.tb00080.x.

5

Brassica napus L: A Case Study

Johannes L. Frieß, Broder Breckling, Kathrin Pascher and Winfried Schröder

Intention and Scope of the Case Study Oilseed Rape (*Brassica napus*)

SPAGE (Self-Propagating Artificial Genetic Element) technologies allow for a proliferation of genetic information on the population level at a higher rate than usual Mendelian inheritance. Currently projected developments of SPAGE mainly aim at a reduction or suppression of animal populations which are considered to be harmful or undesirable (Oye et al. 2014). However, the application of SPAGE is not limited to animals only. In principle, also plant populations can be targeted (National Academies of Sciences 2016). The GeneTip case study on oilseed rape (*Brassica napus*) is intended to assess, which interactions play a role in a plant-specific context to address relevant ecological interactions that need to be fully explored in order to estimate potential risks. For such an assessment, oilseed rape is of particular interest though no application developments are known to be currently on the way. Apart from that, first experiments for instance increased shatter resistance to avoid seed loss during harvest are already in development (Braatz et al. 2017).

Brassica napus exhibits relatively well studied environmental interaction types that are prototypic for many other plants, in particular addressing relations of cultivation and genetic exchange with feral and weedy species (Landbo et al. 1996;

B. Breckling · W. Schröder
Chair of Landscape Ecology, University of Vechta, Vechta, Germany

J. L. Frieß (✉)
Institute of Safety/Security and Risk Sciences, University of Natural Resources and Life Sciences (BOKU), Vienna, Austria
e-mail: johannesfriess@gmx.net

K. Pascher
Department of Integrative Biology and Biodiversity Research, Zoology, University of Natural Resources and Life Sciences, Vienna, Austria

Pascher et al. 2006, 2010, 2017). It is one of the species that has its centre of origin in Europe, and thus a specific responsibility of the European Union to secure sustainability conditions for related wild species may be implied. An analysis of the environmental network of oilseed rape helps to understand the context that is of comparable importance with the prerequisites for various other plant species.

Oilseed rape exemplifies the following risk relevant relations:

The plant is a main crop in Central European areas with significant economic value. It is traded world-wide to a relevant economic extent. Oilseed rape has a variety of technical uses including biodiesel or lubricant for industrial machinery (Moser et al. 2013). It is most widely applied for food (e.g. edible oil, honey) as well as in animal feed (Sarwar et al. 2013). The pollen of the plant is dispersed over very large distances by wind and insects, in rare cases even up to 26 km (Devaux et al. 2005; Ramsay et al. 2003). Not only does oilseed rape frequently grow as a volunteer and establish feral populations, it also forms seed banks, which are viable for up to 15–20 years (D'Hertefeldt et al. 2008; Lutman 1993; Schlink 1998a, b). Its far reaching spatio-temporal spread and persistence is further coupled with a remarkable extent of hybridisation potential to a large number of related species within the *Brassica* genus and partly even to other genera such as *Raphanus*, *Sinapis* and *Erucastrum* (Chèvre et al. 2004). Gene flow is rather well studied and models on the population dynamics exist (Colbach et al. 2001a, b; Habekotté 1997a; Middelhoff et al. 2011). Furthermore, extensive studies on the occurrence of transgene escapees from genetically modified (GM) oilseed rape cultivation and in countries without cultivation of GM oilseed rape have been conducted (Knispel and McLachlan 2010; Simard et al. 2005; Warwick et al. 2003, 2008; Yoshimura et al. 2006). It was also shown, that a self-organised formation of gene stacking can occur in the wild, accumulating unassessed combinations of transgenes even within a very short period of time (Hall et al. 2000; Warwick et al. 2008). Beyond that, there is a still largely unknown hybridization network to other members of the Brassicaceae family, involving bridge species (Eschmann-Grupe et al. 2003; Sobrino-Vesperinas 1988). From invasion biology it is known that plants can undergo an adaptation phase while they persist unrecognised before they expand in range and frequency (cp. Prentis et al. 2008).

The intention of this case study is to consider and accumulate potential cause-effect pathways to be studied in a plant context, which provides some differences compared to animal application of SPAGE techniques.

Oilseed Rape—Biological and Ecological Characteristics

Brassica napus is an allotetraploid combination of its two diploid parent species *Brassica rapa* and *Brassica oleracea*. As both wild parent species used to appear on the coasts of the Atlantic Ocean and the North Sea, several records suggest that amphiploid oilseed rape forms derived from the crossbreeding of *B. rapa* and *B. oleracea* have occurred several times at different sites and with different forms of diploid parents (Song et al. 1990). Comparative genome analyses indicate that the diploid species *B. nigra*, *B. rapa* and *B. oleracea* are derived from hexaploid ancestors.

The oldest archaeological documented findings of oilseed rape in Europe date back to the thirteenth century. There is no evidence for a wild parental species of the crop (Chalhoub et al. 2014). It is assumed that *Brassica napus* has its origin in the Mediterranean, the common distribution area resulting in a *B. napus* geno- and phenotype have been exemplified (Körber-Grohne 1995). Due to its origin, *B. napus* is well adapted to the Mediterranean and central European climate conditions.

Although oilseed rape has characteristics of wild plants such as seed dormancy and adaptation of seed germination capacity to the annual cycle, it does not exist as a wild plant (Ammann and Vogel 1999). Janchen (1972), Adler et al. (1993) describe oilseed rape as largely cultivation dependent, although long-term persistence of feral populations has already been confirmed in several European countries (France: e.g. Pessel et al. 2001; Great Britain: e.g. Crawley and Brown 2004; Germany: e.g. Menzel 2006; Austria: e.g. Pascher et al. 2010; Netherlands: e.g. Tamis and De Jong 2010).

Besides feral oilseed rape which grows on landfills, along roadsides, and in ruderal sites, the so-called volunteer oilseed rape grows as a weed in fields in subsequent crop rotations. This is due to the oilseed rape's secondary dormancy allowing it to remain in the seed banks for multiple years. Up to twelve volunteer oilseed rape plants per 21.6 m^2 were found by (Förster et al. 1998) on former oilseed rape fields in subsequent crops. Thus, volunteers represent an essential factor which needs to be considered in risk assessment of transgenic crop varieties (Pekrun et al. 1998). For instance, 70% of the oilseed rape seeds stored in the soil were still able to germinate after 1.5 years and up to 58% even after five years (Schlink 1998a, b, 1994). Due to the secondary dormancy, germination capacity can even be maintained for more than 10 years in deeper soil layers. Such high survival rates are otherwise achieved only by weed seeds (Mayer et al. 1995).

Oilseed rape as a di-genomic species possesses the complete genomes of its two parental species (Fig. 5.1). For this reason, various interspecific crosses with the parental species proved to be successful in the past (Chen et al. 1988; Gland 1982). Hybridisation of *Brassica napus* subsp. *napus* with other closely related species has been demonstrated several times (e.g. Kerlan et al. 1992; OECD 1997; Scheffler and Dale 1994). In order to successfully hybridise two species, the polyploidy level of the female plant has to be at least as high as that of the male pollen donating plant. Therefore, oilseed rape as a tetraploid plant often performs better as a pollen acceptor than as a pollen donor in hybridisation events with related diploid species (Harberd and McArthur 1980; Sikka 1940).

Allotetraploid hybridisation due to the double genome of oilseed rape increases the genetic variability which can occur in the resulting hybrids. Due to backcrossing with the parent species, new genes can be incorporated into the gene pool (introgression).

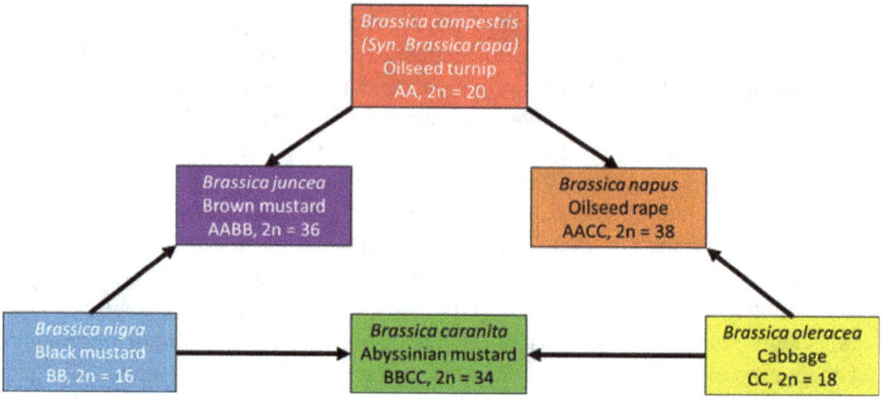

Fig. 5.1 Relations of selected diploid and alloploid members of the genus Brassica. Adapted from Nagaharu 1935

Profile: Oilseed rape, *Brassica napus* L. (Fig. 5.2).
Location: On agricultural land: under cultivation, and as a volunteer in subsequent crop rotations. As feral plants on landfills, along roadsides, on fallow land, former rapeseed fields. Mainly on fresh, nutrient- and alkaline sandy or loamy soils, deep soil with sufficient water supply.
Life form: annual to perennial.
Flower: In cultivation, the main flowering time of feral oilseed rape is between March/April to June. Outside cultivation, flowering extends to October, sometimes even later. Flowering usually lasts for about two to three weeks.
Fertilisation: Oilseed rape is self-compatible and high levels of self-pollination (up to 70–80%) were described. Its brightly coloured petals, productive nectaries, the fragrance, the protogyny and the outwardly open anthers already indicate its potential for cross-pollination, which is conferred by insects; wind pollination also occurs and is responsible for pollen distributions over large distances up to 26 km (Ramsey et al. 2003).
Fruit: The fruits are up to 8 cm long and up to 4 mm thick, with a 1–2 cm long beak. At the time of the harvest, there are about 30–50 plants per square meter in the field. Each plant produces 7–10 lateral shoots and 120–200 pods.
Seeds: Each pod contains about 18–30 small, one millimetre large, spherical seeds. Plants produce more than 1000–1400 seeds. Thousand grain weight (TGW) is approx. 5 g.
Germination: Germination is rapid and uniform. Seeds survive in the soil by secondary dormancy varies depending on the germination temperature, the light supply, the variety and the season. *B. napus* is frost resistant down to about −15 to −20 °C.
Phenotypic plasticity: Depending on exposition, climatic conditions and nutrient supply, plants can considerably vary in height, branching and number of pods (between one single up to several hundreds.

(profile adapted from Breckling et al. 2004).

Fig. 5.2 Flowering oilseed rape

Production, Uses and Genetic Modification

Oilseed rape is one of the first crops of which genetically modified varieties were grown on a large scale. Multiple varieties with various properties, in particular herbicide resistance have been developed. Genetically modified oilseed rape varieties have been grown in the USA and Canada since 1996 (ISAAA 2017), and in Australia since 2008 (James 2015). In the EU, albeit there have been 383 field trials in 11 countries,[1] there are so far only approvals for import of GM oilseed rape as food and feed, but not for commercial cultivation inside the EU. Table 5.1 lists the top rapeseed producing countries concerning the mass and area of cultivated rapeseed. If known, the estimated percentage of GM rapeseed is given.

Up until the 1970s, due to its high erucic acid and glucosinolate contents, oilseed rape could hardly be used for consumption. Since 1974, however, LEAR-varieties (low erucic acid rapeseed) with an erucic acid content below 2% became available. Developed in Canada, double low varieties with low erucic acid and almost no glucosinolate contents were bred by traditional means (canola). This allowed the increased use of the oil and seed protein feed for non-ruminating animals such as pig and chicken. Next to the widely cultivated double low varieties, HEAR—(high erucic acid rapeseed), HO—(high oleic acid), HOLLi—(High oleic, low linoleic), and triple-zero varieties exist. The latter in addition to the low erucic acid and glucosinolate contents, is also low in fibre content.

[1] https://gmoinfo.jrc.ec.europa.eu/overview-plants.aspx.

Table 5.1 Top oilseed rape producers and cultivation areas in 2016 (FAOSTAT, ISAAA, USDA)

Country	Produced rapeseed /tonnes	Cultivated area /ha	GMO Percentage
Canada	18,423,600	7,990,300	93%
China	15,281,624	7,614,543	
India	6,797,000	5,762,000	
France	4,727,961	1,550,720	
Germany	4,579,600	1,325,700	
Australia	2,944,000	2,357,000	23%
Poland	2,219,270	826,946	
United Kingdom	1,775,000	579,000	
United States	1,403,650	686,440	90%
Czech Republic	1,359,125	392,991	
Romania	1,292,779	455,048	
Ukraine	1,153,910	449,930	
Russia	998,932	911,849	
Hungary	608,700	222,085	

Oilseed rape has a wide range of applications such as cooking oil, ingredient of cosmetics but also pesticides, biofuels, etc. It is the third-largest used source of vegetable oil in the world (USDA 2018) followed by soybeans and oil palms (Kumar et al. 2007; Langhof and Rühl 2017).

Potential Hybridisation Partners

Besides oilseed rape's capability to persist in the seed bank and re-emerge either as feral or volunteer, the plant also has multiple potential hybridisation partners. With decreasing likelihood of hybridisation according to a study in Austria (Pascher et al. 2000), these are:

Turnip	*Brassica rapa* (feral): partental species
Cabbage	*Brassica oleracea* (feral): parental species
Longstalked rape	*Brassica elongata*
Brown mustard	*Brassica juncea*
Black mustard	*Brassica nigra*
Haddick	*Raphanus raphanistrum*
Wild mustard	*Sinapis arvensis*
White mustard	*Sinapis alba*
Perennial wall-rocket	*Diplotaxis tenuifolia*
Annual wall-rocket	*Diplotaxis muralis*
Radish	*Raphanus sativus*

Garden rocket	*Eruca sativa*
Watercress dogmustard	*Erucastrum nasturtiifolium*
Hairy rocket	*Erucastrum gallicum*
Steppe cabbage	*Rapistrum perenne*
Shortpod mustard	*Hirschfeldia incana* (formerly *Brassica geniculata*)
Annual bastardcabbage	*Rapistrum rugosum*
Austrian hare's ear mustard	*Coringia austriaca*
Oriental hare's ear mustard	*Conringia orientalis*
Tatarian cabbage	*Crambe tataria* (endangered species in Austria, only one location of occurence)
Hedge mustard	*Sisymbrium* spp

Additional species for single European countries are listed in Chévre et al. (2004). A study from Belgium (Devos et al. 2009) established a gene flow index to evaluate the introgessive hybridisation potential (IHP) of each relative. They conclude that *Brassica rapa* as one of the parents of oilseed rape has the highest introgressive hybridisation propensity (IHP value = 11.5), followed by *Hirschfeldia incana* and *Raphanus raphanistrum* (IHP = 6.7), *Brassica juncea* (IHP = 5.1), *Diplotaxis tenuifolia* and *Sinapis arvensis* (IHP = 4.5), in Flanders.

The following paragraphs take a closer look at some of the well-documented hybridisation partners of oilseed rape most common in Germany.

Oilseed turnip (Brassica rapa)

Brassica rapa is not only a highly developed crop but also one of the oldest crops that was grown by man. *B. rapa ssp campestris* is the wild form of todays cultivated turnip, a breeding source of a variety of crops. Compared to oilseed rape, turnip (Fig. 5.3) has some morphological differences (Fischbeck et al. 1982). Since *B. napus* (AACC) is an allotetraploid descendant of *B. rapa* (AA) and *B. oleracea* (CC), spontaneous interspecific hybridisation between oilseed rape and turnip occurs frequently. Spontaneous crosses have been observed under natural conditions, among others by Stace (2010). Successful hybridisation has also been proven under controlled as well as under field conditions in different countries: Canada, England, Denmark, New Zealand, Australia, Czech Republic, (Bing et al. 1991; Jørgensen et al. 1996; Jørgensen and Andersen 1994; Metz et al. 1997; Mikkelsen et al. 1996; Scott and Wilkinson 1998).

Hybridisation is most successful when oilseed rape functions as the female parent (Scheffler and Dale 1994), but crossings in both directions are possible (Becker 1951). Jørgensen and Andersen (1994) could show an interspecific hybridisation rate of 9–93% in crossbreeding of *Brassica napus* and *Brassica rapa*. In hybridisation experiments between oilseed rape and turnip it was found that all F_1-hybrids were morphologically similar to oilseed rape, but showed a reduced pollen fertility of around 55% (Warwick et al. 2003). The introgression rate of herbicide resistance genes from oilseed rape to turnip was observed to be low (Hansen et al. 2001; Jørgensen 1999; Norris and Sweet 2002). Interspecific hybrids are able to backcross with *B. rapa* as a female parent (Jørgensen et al. 1996; Mikkelsen et al. 1996).

Fig. 5.3 As one of the most significant hybridization partners, *Brassica rapa* also occurs outside cultivation at feral sites (Vechta 2019)

Haddick (*Raphanus raphanistrum*)

The haddick, also called wild radish, is native to Europe, Western Asia and parts of Northern Africa. It is considered a wide-spread and invasive weed that grows in summer crops, former oilseed rape fields but also in urban environments and along roads. *Raphanus raphanistrum* shows a great morphological variability. Seeds of haddick remain viable in the soil for a long time due to a restistant seed coat (Darmency et al. 1998).

Already in 1924, intergeneric hybrids between the more distantly related genera *Raphanus* and *Brassica* have been successfully produced (Karpechenko 1924). Transgenic oilseed rape hybrids can also be created by ovary culture (Kerlan et al. 1993). Under natural conditions successful hybridisation was detected between male sterile specimens of *B. napus* and *R. raphanistrum* (Eber et al. 1994). In that experiment, frequent chromosome pairing and the presence of multivalent compounds have indicated that recombination is also possible between the chromosomes of different genomes. The recombination between the genomes of *B. napus* and *R. raphanistrum* could also be shown experimentally (Baranger et al. 1995).

Spontaneous hybridisation of oilseed rape and haddick, even under natural field conditions, was achieved at a relatively high frequency in France (Darmency et al. 1998) and Australia (Rieger et al. 2001). Intergeneric gene flow may occur mainly through introgression of the transgene into the genome of the weeds (Chèvre et al. 1997).

The hybrid frequency is expected to range between 0.006 and 0.2% of the total seeds produced (Darmency et al. 1998). In male sterile oilseed rape plants surrounded by haddick, even up to 37 hybrids per plant could be detected. Seed production of the F_1 and F_2 generations even reached rates of 0.4–2% in comparison to wild haddick.

In contrast, Australian studies on gene flow between oilseed rape and haddick showed low hybridisation rates ($<4 \times 10^{-8}$). When haddick acted as the maternal partner, no hybrids were detected (Rieger et al. 2001).

Perennial wall-rocket (*Diplotaxis tenuifolia*) and Annual wall-rocket (*Diplotaxis muralis*).

The genus *Diplotaxis* encompasses around 30 species. *Diplotaxis* is spread across the warmer West Eurasian as well as in the East African mountains. Its centre of diversity lies in the south-western Mediterranean area (Eschmann-Grupe et al. 2003).

Diplotaxis tenuifolia occurs in moderately dry ruderal areas, which can be found especially in harbour areas, on railways or increasingly in urban areas but also in farmland. It is one of the four species in Austria with the highest hybridisation probability with oilseed rape due to the close genetic relationship and the frequent occurrence (Pascher et al. 2000). Hybridisation events with oilseed rape were reported by Ringdahl et al. (1987). Similar to *Diplotaxis muralis*, hybridisation with oilseed rape as a female parent was unsuccessful (Salisbury 1989). The crossing between *D. muralis* and *D. tenuifolia* produced a successful F_1-generation (Eschmann-Grupe et al. 2003; Sobrino-Vesperinas 1988), so hybridisations with *Diplotaxis sp.* and oilseed rape might also create viable hybrids via such “bridges”.

D. muralis prefers moderately dry soils and can be found on short-lived weed fields and, as the name implies, close to walls. In a crossing experiment by Ringdahl et al. (1987), a total of 285 flowers of 9 *Diplotaxis* plants were pollinated with oilseed rape pollen. 157 pods (60.1%) with 607 seeds were formed, 31 (5.1%) of these seeds were hybrids(Ringdahl et al. 1987). In Australia, *Diplotaxis muralis* produces viable but sterile hybrids after forced pollination with oilseed rape pollen. The cross yielded 0.054 seeds per pollinated flower (Salisbury 1991). In contrast to the results of (Salisbury 1989) attempts of hybridisation with oilseed rape as a female parent produced hybrids in studies by Bijral and Sharma (1996).

Conclusions on Gene Flow Potential

There are numerous, widely distributed hybridisation partners with a high level of cross-fertilisation within the Brassicaceae family—almost exclusively among the species of the tribe Brassiceae—which makes the formation of intra- and interspecific hybrids quite common (Chèvre et al. 2004). Large-scale cultivation of transgenic oilseed rape over several years supports the formation of hybrids and thus the propagation of transgenes. It can establish feral populations at ruderal sites and occasionally even in natural habitats such as river banks (Pascher et al. 2000, 2017; Pascher and Gollmann 1999). Also, the persistence of oilseed rape seeds in the soil is an essential factor. There are various studies with different results concerning the distribution of pollen or hybridisation frequencies of related crossing partners of oilseed rape. For instance, the outcrossing rates of the same plant species differ by several orders of magnitude in different experiments. Furthermore, the frequency of naturally occurring gene transfer varies to a large extent not only between plant families and species, but also between populations, individuals and even from year to year, hinting at the importance of the current environmental conditions such as temperatures and moisture during 'pollen rain', wind force and direction, size of donor and receiver population, distance etc. or different test setup (e.g. Gliddon 1999).

Finally, knowledge about distribution, abundance and flowering times of single relevant species are indispensable. In the GenEERA project (2001–2004), the cultivation sites of oilseed rape and the co-occurrences of potential hybridisation partners in the vicinity was assessed and quantified in maps and geostatistically evaluated for typical Northern German landscapes and an urban area, the city of Bremen. Figure 5.4a shows the cultivation situation. Figure 5.4b indicates the approx. 500 km^2 that were visited in all publicly accessible areas to map the number of oilseed rape single plants and populations as well as its potential hybridisation partners.

Pests and Pathogens

As the *Brassica* species are important native components of many ecosystems in temperate climate, they provide forage for a large number of invertebrates and several other organisms such as mice and birds (Organisation for Economic Co-operation and Development OECD 2012). Herbivory is an important limiting factor in the commercial crop production of *Brassica* species (Kimber and McGregor 1995; Lamb 1989).

Oilseed rape is also targeted by a wide range of pathogens, bacterial, fungal, viral, phytoplasmal, and miscellaneous other diseases. Three diseases are particularly dangerous, because they are pandemic and have a potential for major crop injury. These fungal diseases are blackleg, also known as stem canker, Sclerotinia stem rot and clubroot. The conventional development of resistant varieties is difficult due to the multi race pathogenicity (Organisation for Economic Co-operation and Development OECD 2012).

Pollen Transfer and Gene Flow

Oilseed rape fields produce pollen for about three weeks, depending on the particular weather conditions. *Brassica napus* is self-fertile. But since the stigma is mature around three days before and after the anther, cross-pollination is also well possible. Therefore, seed formation mainly consists of a mixture of cross- and self-pollination of variable ratios. The ratio of cross fertilisation depends on various factors such as the weather at the time of flowering and the genetic predisposition of the lineages

(a)

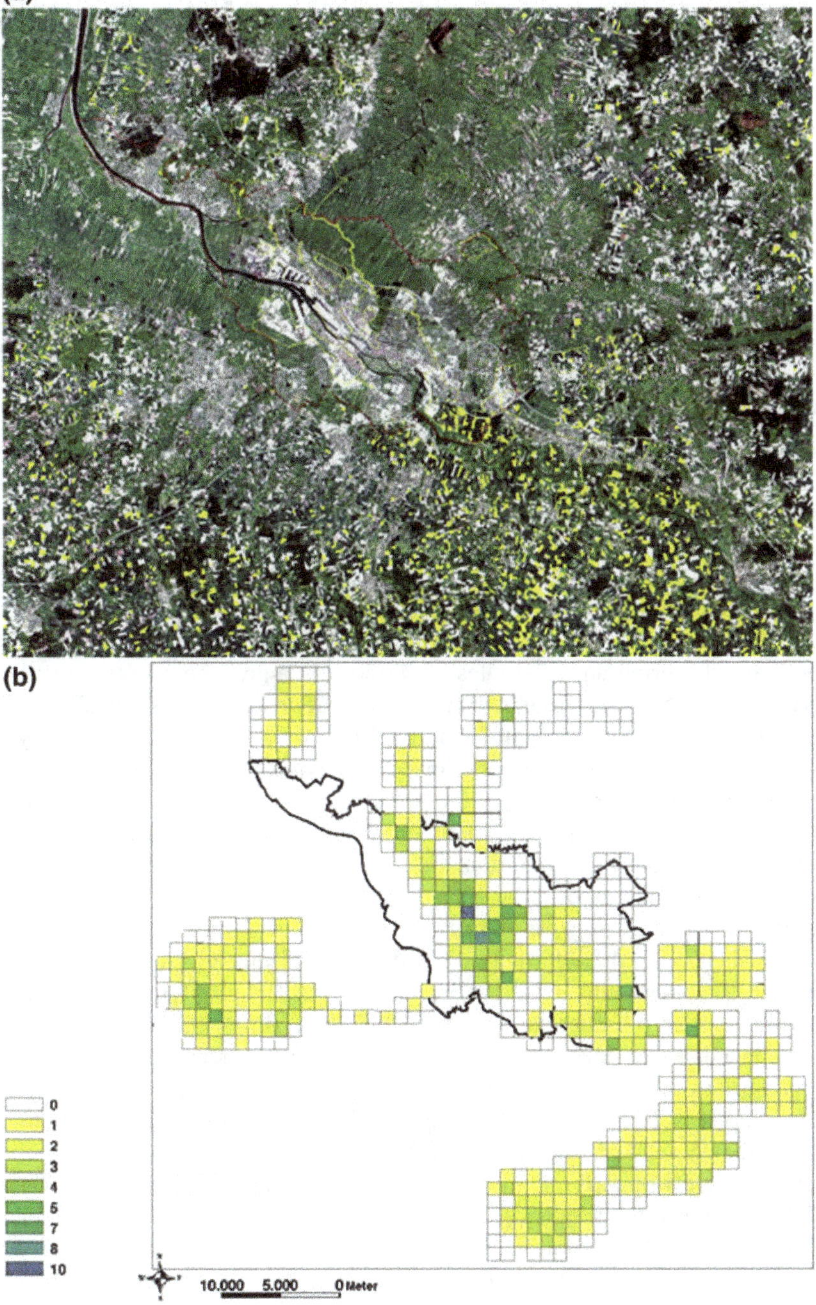

◂**Fig. 5.4** **a** Satellite image of the surroundings of Bremen (Landsat) from May 2001. Oilseed rape fields are marked in yellow. The border of Bremen is indicated as a thin red line. Areas visited during the field survey are surrounded by a yellow line. From: (Menzel and Born 2004, p. 43). **b** Number of hybridisation partners mapped during the field survey in 2001. Solid Black line: Bremen city limit. Light grey lines: Main traffic axes. The colour code refers to the number of species (oilseed rape and hybridisation partners) found in the respective grid element (1 km^2). It is apparent that the central urban region harbours the highest species number of the considered plant group. From: Menzel and Born (2004, p. 31)

(Hühn and Rakow 1979). Exact details of the quantitative effect of individual factors are not available in the literature and the estimates for self- and cross-pollination vary greatly. For instance, Hühn and Rakow (1979) assume that self-pollination accounts for an average of two-thirds of oilseed rape. Timmons et al. (1995) reported outcrossing rates of 5–55% based on various data from the literature under field conditions. Neemann and Scherwaß (1999) assume an average cross-fertilisation rate of 20–30%. Other authors report the outcrossing rate in rapeseed to be 5–30% (Rakow and Woods 1987), 22–36% (Scheffler et al. 1993) or 41% (Lavigne et al. 1998).

Oilseed rape pollen is more or less spherical with a diameter of 25 μm. Their size corresponds approximately to the size of many fungal spores, which are distributed primarily by wind (McCartney and Lacey 1991). A yield reduction of 33–50% in oilseed rape flowering under windless conditions in a greenhouse experiment suggests that wind also plays a role in the pollination of oilseed rape, especially for long distance dispersal (Timmons et al. 1995). Furthermore, it is reported that good yields can be achieved even in the absence of insects (Wilkinson et al. 1995). The attractive flowers of oilseed rape are mainly pollinated by insects, honey bees play a significant role in pollination (Gerdemann-Knörck and Tegeder 1997). To secure purity of the varieties in seed production, isolation distances of 100 m for certified seed and 200 m for basic seed are required in the EU (Ingram 2000). In Austria, although several international scientists suggest an isolation distance of 4 km considering 26 km as maximum investigated confirmed pollen dispersal distance (Ramsey et al. 2003), isolation distances of only 200 m for lineage varieties and 500 m for hybrid varieties are put in practice (Pascher and Dolezel 2005). The aim of this measure is not an absolute avoidance of gene flow, but a limitation as efficient as possible.

Pollen transport depends on various factors such as the respective lineage, the location and season (Raybould and Gray 1993). Also, seed spillage is a significant factor, since on the one hand large amounts of seed loss occur during harvest of oilseed rape are lost in the field [30–40 kg/ha (Gerdemann-Knörck and Tegeder 1997)]. Because of the small seed size seeds are regularly spilled during transport and handling activities (Adolphi 1995; Crawley et al. 1993; Franzaring et al. 2008; Pascher et al. 2017; Warwick 1997).

These circumstances lead to an almost ubiquitous occurrence of oilseed rape in the farmland and to the situation that oilseed rape plants combine the characteristics of different oilseed rape lineages or varieties. For genetically modified varieties, (Hall et al. 2000) described the occurrence of plant individuals with multiple herbicide resistances in a location where varieties with different single resistances were grown. In this case, the triple resistant individuals were detected up to a distance of 550 m from the putative pollen source.

Gene Flow by Airborne Pollen Transport

Although information on the spread of oilseed rape pollen exists, there are no quantitative data on the proportions or the respective influencing factors available in the literature. However, a large number of influencing factors (Ingram 2000; Treu and Emberlin 2000) as well as optimal conditions for high and low outcrossing rates (Feil and Schmid 2001) are compiled. It is estimated that under typical weather conditions in Britain (wind speed of 2 m/s, with convection currents), a potential pollen drift of almost 172.8 km is possible within 24 h (7.2 km/h). Even a pollen transport over 864 km would be possible within one day with wind speeds of 10 m/s. Therefore, it is not surprising that pollen can be detected in the air above the middle of the Atlantic Ocean (Treu and Emberlin 2000). However, only a small amount of the released pollen drifts and no quantitative data on the long distance spread of pollen is available. Factors influencing the quantity of pollen accredited to regional gene flow may be the type of plant and variety characteristics, the current weather conditions (particularly wind conditions) and the time of pollination, especially the time of the day. For example, at night, there are different conditions for pollination than during the day. At night, the long-distance transmission of pollen (regional gene flow) is usually lower (McCartney and Lacey 1991).

Furthermore, the landscape configuration has a dominant influence on the extent of local gene flow between crops and wild relatives (Squire et al. 1999). The rate of cross-breeding from more distant sources is higher when there are no local pollen sources. The highest pollen concentrations occur on days with intense insolation and strong wind (McCartney and Lacey 1991). Correspondingly, a reduction in the amount of detected oilseed rape pollen to 1.4% compared to the previous year was attributed to heavy rainfalls and high humidity in the second year (Scott 1970 referenced by Treu and Emberlin 2000). With the two factors humidity and more significantly the wind, a correlation to the pollen concentration but not to the relation between wind speed and distance could be found. Instead, it is suspected that upward air movements are crucial for the pollens' trajectory (compare viable maize pollen in 1.8 km height), while wind speeds play a minor role (Feil and Schmid 2001). Moreover, the main wind direction plays a major role. Messeguer et al. (2006) assessed the influence of wind for maize cross-pollination.

Lineage-Specific Factors of Actual Gene Flow

Beside the factors that influence potential gene flow, the actual entry of pollen into plant populations (actual gene flow via pollen) are influenced by additional factors. These factors are also dependent on the oilseed rape lineage such as flowering periods of donor and receptor populations, outcrossing rates, self- and cross-fertilisation rates, hybridisation rates, pollen life-span and morphological characteristics regarding the compatibility of pollen and stigma.

Outcrossing rates observed in fertile oilseed rape are not transferable to male sterile oilseed rape used for hybrid seed production. Conversely, observations with male sterile recipients or those incapable of self-fertilisation cannot be transferred to male fertile of self-pollinating populations (Ingram 2000).

Hybrid varieties, such as Synergy, appear to be particularly prone to introgression (Simpson et al. 1999), as they worked with male sterile plants and the cultivated crop regularly contained male sterile plants. However, in some hybrid varieties, fertility in the growing crop is mostly restored (not always completely), so foreign pollen in turn has to compete with local pollen (Ingram 2000).

Not-Lineage-Specific Factors of Actual Gene Flow

The following factors influence the competitive situation between the pollen of the receptor population and 'foreign' pollen. This competitive relationship is a significant factor which pollen fertilises the flowers of the receptor population. These factors are the geographic distance to plants and receptor populations, crop density and size ratio of donor to receptor populations (Feil and Schmid 2001) as well as competition between own and foreign pollen arrived at the stigma and growing towards the ovary.

Gene Flow via Pollen Transport by Insects

Insects play the most important role in the fertilisation of oilseed rape. Some farmers even pay beekeepers to set up beehives near their fields. This is of special importance for sterile hybrid crops (Ramsey et al. 1999) which, for instance, account for approximately 75% of the oilseed rape varieties in Austria. All of the GM oilseed rape varieties are hybrids. The relatively high sugar concentration in the nectar of the plant attracts bees and other insects over long distances (Ramsey et al. 1999; Saure et al. 1999a, b). Honey bees provide 90% of insect pollination in oilseed rape (Mesquida et al. 1988). However, solitary bees and bumblebees, if they occur, are comparably efficient in pollination. Other species such as dipterans, butterflies and beetles occur less frequently and have a lower pollination performance (Mesquida et al. 1988; Saure et al. 1999a, b).

Information on insect flight distances vary greatly. The data on the flight distances of honey bees range from 600 m (Osborne et al. 1999) over 2 or 5 km (according to beekeepers) and 2.75 and 4.6 miles (4.42 and 7.4 km) (Eckert 1933) up to 14 km (Saure et al. 1999a, b). Clustering radii of up to 6 km, which would correspond to an area of up to 100 km^2 around a hive were observed (Waddington et al. 1994). For solitary bees with a small body size, flight distances of 200 m, up to 800 m for sand bees, and 2 km for bumblebees were measured (Saure et al. 1999a, b). Bees can carry up to 60,000 oilseed rape pollen on their body in addition to the pollen in their pollen sacs. In the hive, worker bees clean each other. Thereby, pollen transfer to other bees takes place. Thus, bees leaving the hive probably carry a mixture of fertile pollen on their bodies (Ramsey et al. 1999).

In summary, the high variability concerning data on the flight activity of insects also have an unpredictable influence on gene flow. Insect activity is reduced during cold, wet, cloudy and stormy weather. Reduced insect activity is also observed during an especially hot and dry year (Brown and Brown 1996).

The extrapolation from small- to large-scale experiments is considered as difficult. For instance, a small pollen source experiment extrapolated 2–11% of the original pollen level 100 m from the pollen source. But a large-scale study showed pollen densities of 27–69% at 100 m distance from an oilseed rape field (Timmons et al. 1995; Wilkinson et al. 1995). Therefore, the extrapolation from small pollen sources to the field level is problematic (Wilkinson et al. 1995). However, small-scale experiments yield valuable results, if the gene flow from or into small stocks is considered. Prime examples for this would be feral oilseed rape populations also descending from transport losses as well as volunteer populations.

Seed Persistence and Germination in Oilseed Rape

Apart from *B. napus* capability to spread its genetic material by pollen, also volunteers acting as transgene donors due to seed persistence have to be considered. It is argued that the risk of spread of e.g. the *pat*-gene (glufosinate resistance gene) by threshing losses at harvest and volunteering in the next cultivation cycle is much higher than by pollen transport (Fischbeck 1998). Therefore, the seed persistence of oilseed rape plays a major role in its gene flow pattern. Thereby, not only the persistence over time but also the spatial distribution of seeds has to be taken into account.

In oilseed rape fields, on average 200–300 kg of grain per hectare remain after harvest (Pekrun et al. 1998). Whereas, the optimal sowing rates are as low as 60–90 seeds/m^2 (Männer 2000). In Northern Germany, however, occurrence of 400 volunteers/m^2 is not uncommon (Gerdemann-Knörck and Tegeder 1997). In Southern Germany, 0–202 seeds/m^2 were observed one to six years after oilseed rape cultivation. Whereby, seed counts above 100 occurred only at locations, where unusually large quantities of lost oilseed rape were left on the field due to late harvesting or hail (Roller et al. 2001). Other studies even report 10,000 lost seeds/m^2 (Lutman 1993). This amount of persistent seeds has an apparent impact on volunteer emergence.

Norris et al. (1999) observed that in the United Kingdom, the number of volunteer plants can vary greatly from one area to another. This depends on details of the agricultural practice.

For instance, it was observed that an erucic acid content of less than two percent was achieved only seven years after a switch to an erucic acid free oilseed rape variety (Sauremann 1987) referenced by (Schlink 1994). Also, a growth of oilseed rape varieties cultivated ten years ago could be found (Röbbelen 1986) referenced by Schlink (1994). It could be shown in experiments that after ten years on average 0.5% of the buried seeds survived. The maximum survival was 4.7%. Of those persistent seeds, over 96% sprouted and developed viable seedlings within four days after being exposed to daylight on the soil surface (Schlink 1998b).

These high persistence probabilities of above 70% over a one-and-a-half-year period and nearly 60% after five years in the soil resemble those of weed seeds. Also a change in the germination rate in the annual cycle is more characteristic for wild plants (Pekrun et al. 1997). Due to the traits of his two wild parental species, oilseed rape has maintained 'weedy' characteristics (Schlink 1994).

The long-term survival of *B. napus* seeds and resulting volunteers is attributable to dormancy. Dormancy grants seeds the ability to survive for extended periods of time in the soil also under changing environmental conditions. Seeds are defined to be dormant, if they do not germinate under the optimum environmental conditions and an additional factor or environmental stimulus is required for germination. There is a distinction between primary and secondary dormancy. The primary dormancy prevents germinations on the mother plant and up to a certain period of time after the seed has dropped. Secondary dormancy on the other hand, is induced by unfavourable environmental conditions on the swollen seeds and can either be broken by specific environmental stimuli or degraded over a longer time period. Thereby, a clear distinction between primary and second dormancy is not always possible and dormancy should not be confused with quiescence. Quiescence happens, when seeds do not germinate as a result of adverse environmental conditions e.g. lack of water, air or due to unfavourable temperatures. Influencing factors for germination and survival are listed in Table 5.2.

Genetic Modifications in Oilseed Rape

Oilseed rape is one of the first crops that were genetically modified. Several varieties were tested, deliberately released and commercialised. Most of them confer herbicide resistances to a number of active ingredients, among of which glyphosate, glufosinate, bromoxynil and others. Also varieties with altered plant metabolism, in particular fatty acid composition, and male sterility have been created. Moreover, Bt-Insect toxicity (Halfhill et al. 2002) as well as growth form (Gressel 1999; Reuter et al. 2008), and various other traits have been tested so far. A survey of genetic modifications, including oilseed rape is provided by the Food and Agriculture Organisation

Table 5.2 Factors that influence seed survival and germination

Seed survival influencing factors	Seed germination influencing factors
Variety characteristics	Variety characteristics
Year of the seed burial	Time of the year
Storage duration in the soil	Storage duration in the soil
The depth at which the seed is stored in the soil	Water supply
Soil conditions	Soil conditions
Habitat	Temperature
Storage temperature before storage in the soil	Light availability
Duration of stress factors (see Pekrun, Lutman et al. 1997b)	Short-term light exposure while Otherwise in darkness
Amplitude in temperature changes (see Pekrun Lutman et al. 1997b, c)	Stratification
Age of the seeds	Drying

of the United Nations (FAO[2]), by the Biosafety Clearing House mechanism of the Biosafety Convention (Cartagena Protocol[3]) and for deliberate release and placing on the market of GMO by the Joint Research Centre of the European Union.[4]

Persistence of Genetically Modified Oilseed Rape Outside Fields

Breckling and Menzel (2004) discussed the following implications in the light of transgenic oilseed rape persistence. They discuss general implications, effects on agriculture and cultivation implications and effects on biodiversity and wildlife.

General Implications

If no additional fitness differences occur, (Breckling and Menzel 2004) conclude, transgenic oilseed rape would be likely to disperse as widely as conventional varieties with regard to the ecological and physiological characteristics of the species. Wild

[2]https://www.fao.org/food/food-safety-quality/gm-foods-platform/en/, https://www.fao.org/food/food-safety-quality/gm-foods-platform/browse-information-by/commodity/commodity-details/en/?com=38947.

[3]https://bch.cbd.int/database/organisms/default.shtml, https://bch.cbd.int/database/results?searchid=742060.

[4]https://gmoinfo.jrc.ec.europa.eu/Default.aspx, https://gmoinfo.jrc.ec.europa.eu/gmp_browse.aspx.

Fig. 5.5 Escape from cultivation: single oilseed rape plant growing on the margin of a cereal filed in Goldenstedt, Lower Saxony (Germany) 2017

populations of *B. napus* receive a steady input due to seed losses from cultivation, seed transport by vehicles, animals (e.g. wild boar) or man. Buried seeds from the soil seed bank eventually emerge when soil substrate is transported and redeployed. All these factors support sustaining of feral populations. Based on the empirical observation, that urban feral sites are important as centres of occurrence for oilseed rape as well as for potential hybridisation partners (Menzel 2006, see also Fig. 5.5), It is argued that the conditions for genetic interaction (i.e. cross-pollination) are largely provided not only in rural but also in urban environments. The conditions for a fixation of genetic traits seem even more favourable in urban environments due to the smaller population sizes (Klinger and Ellstrand 1994). Since oilseed rape frequently grows along traffic axes and seeds are transported attached to car tires, a rapid translocation seems possible. If an invasive potential should be acquired, cars would strongly contribute to large distance transport. von der Lippe and Kowarik (2007) demonstrated this in an empirical investigation.

Agricultural Implications

Concerning agricultural implications, it is argued that the introduction of transgenic, herbicide resistant varieties would require important changes of the general agricultural practice, for GMO growers and neighbouring conventional fields. Considering the issue of the pesticide treadmill, (Breckling and Menzel 2004) state that due to the fact that harvest seed losses occur on the field to a large extent (Cramer 1990), herbicide resistance will lose efficiency in volunteer control leading to a reduced number of active herbicide compounds available for weed management and volunteer control. They further elaborate that since pollen transfer is efficient over large distances (Treu and Emberlin 2000), cross-pollination-conferred GM-contamination of neighbouring fields regularly occurs. Thus, seed losses in subsequent generations could cause weed management problems, even if the level of contamination may remain below the legal threshold (Pekrun 1994). Tracing the origin of crop contamination to a specific grower might not always be possible, even if the event (the specific variety) and thus the patent holder can be identified through genetic analysis. For Central European cultivation conditions at least, the gain of transgenic herbicide resistances is low (Augustin et al. 1998). As soon as oilseed rape has reached a certain height, it can usually outcompete weeds. Thus, Breckling and Menzel (2004) conclude that in oilseed rape cultivation, weed control before crop germination is usually sufficient and later herbicide treatments are rarely required. This applies to Central Europe; in other regions it may be different. Here, the small advantage in cultivation would have to outweigh the substantial cost for other growers as well as for the general public to cover the management system and documentation and damage regulation cost as well as separate conventional and transgenic processing and commodity flows. This may be one of the reasons, why transgenic varieties are not admitted for cultivation in the EU up to now.

Conservation Implications

As the fixation of transgenes in feral populations appears to be rather probable, conservational issues are implied. In reference to the concept of "genetic swamping" (Snow 2002) which addresses the threat of cultivation-specific, fitness altering genes to wild type populations, Breckling and Menzel (2004) coined the term "transgenic swamping" which refers to transgenes escaping to feral populations. This seems inevitable in the case of oilseed rape. The impact of persisting transgenes on biodiversity is difficult to estimate. However, the invasive potential of released plants has been discussed previously (Hurka et al. 2003) and applications of the precautionary principle are in place also to prevent harm to biodiversity as a protected ecological good. Breckling and Menzel (2004) argue that in the light of the centre of biodiversity of oilseed rape being located in Europe and due to the inevitability of the transgene escape from cultivation to feral populations (Fig. 5.5), an unpredictable high number of genotypes would be anticipated.

Detection of Unintended Spread of Transgenic Oilseed Rape in Various Countries

The following section summarises data collated by Testbiotech (Bauer-Panskus et al. 2013). Testbiotech imposingly evidences the uncontrolled and unintended spread and outcrossing of transgenic oilseed rape varieties for six countries.

Canada

Canada is one of the countries that grows herbicide tolerant oilseed rape in large quantities. Transgenes were found in nearly all of the conventional Canadian oilseed rape seed supply (Friesen et al. 2003). Feral populations had developed at the edges of fields and along roadsides. In the province of Manitoba, 88% of feral oilseed rape populations examined contained glyphosate tolerant plants. About 50% of the plants were tolerant to both herbicides, imidazoline and glufosinate (Knispel et al. 2008). 93 out of 100 feral oilseed rape plants along field edges or roadsides tested, contained transgenic constructs (Knispel and McLachlan 2010). All feral populations tested in another study contained hybrids with *Brassica rapa* (Simard et al. 2006). Tests revealed that nearly no fitness costs are associated with the stacking of transgenes in oilseed rape plants (Simard et al. 2005). Persistence of such hybrid populations over time was affirmed by a long term survey which showed that feral hybrid populations *of B. napus* × *B. rapa*, despite a decreased fitness, persisted over six consecutive years (Warwick et al. 2008). Also transgenic volunteers could be found up to seven years after the original cultivation (Beckie and Warwick 2010).

USA

In the United States, commercial cultivation of genetically engineered oilseed rape started early and, at present, accounts for more than 90% of all oilseed rape fields. Unintended large scale dispersal of herbicide-tolerant oilseed rape along roadsides was demonstrated in North Dakota (Schafer et al. 2011). Out of all oilseed rape plants growing along roadsides, 80% tested positive for genetic modification. Of these plants one half contained the *cp4epsps*-gene for glyphosate tolerance, the other half the *pat*-gene encoding a tolerance to glufosinate. Some plants were tolerant to both herbicides.

Japan

Japan itself does not cultivate but imports genetically modified oilseed rape. The first studies on the presence of transgenic oilseed rape in ruderal habitats in Japan were published in 2005 (Saji et al. 2005). In the proximity of ports like Kashima, Chiba, Nagoya and Kobe as well as along transportation routes to industry plants where oilseed rape seeds are processed, plants were found that proved to be resistant to glyphosate or glufosinate although these transgenic varieties have never been cultivated. Feral oilseed rape populations co-occur with wild populations of *B. juncea* in port areas (Kawata et al. 2009). Transgenic oilseed rape plants were detected that had hybridised with each other and consequently, were tolerant to both herbicides (Aono et al. 2006). In follow-up studies feral populations were found along further transportation routes (Nishizawa et al. 2009) and in areas close to all other major ports (such as Shimizu, Yokkaichi, Mizushima, Hakata, or Fukushima) (see for example Kawata et al. (2009). Mizuguti et al. (2011) came to the conclusion that the investigated oilseed rape populations were able to self-sustain over time. In 2008 in the proximity of Yokkaichi port, 90% of all tested plants proved to be genetically modified. Also in that area, the first transgenic hybrids between *B. napus* and *B. rapa* were found (Aono et al. 2011). Under the influence of climatic conditions, the properties of feral transgenic oilseed rape plants might have changed. Plants with greater height that have also become perennial were found (Kawata et al. 2009), whereas oilseed rape and all other *Brassica* species growing in Japan are usually annual.

Australia

In Western Australia, herbicide tolerant oilseed rape is cultivated only in certain areas and makes up less than ten percent of the overall oilseed rape acreage (McCauley et al. 2012). Nevertheless, more than 60% of the samples taken from feral oilseed rape populations tested positive for glyphosate tolerance (Conservation Council of Western Australia CCWA 2012).

European Union

Although transgenic oilseed rape was never admitted for cultivation in theEuropean Union, several field trials have taken place, the "transgen" website lists 383 in 11 countries.[5] In Germany (North Rhine-Westphalia) feral genetically modified oilseed rape was found 700 m from a former trial field (Hofmann et al. 2007). Even ten years after field trials, transgenic oilseed rape was found in Sweden (D'Hertefeldt et al. 2008). In Austria, feral conventional oilseed rape single plants up to large populations as well as related wild hybridisation partners were regularly identified along transportation routes, border railway stations, railway stations, at switchyards, ports and oil mills (Pascher et al. 2017). Feral populations were also observed in France (Garnier et al. 2008; Pivard et al. 2008), the Netherlands (Tamis and De Jong 2010) and Great Britain (Crawley and Brown 2004; Squire et al. 2011).

Switzerland

Fifty out of 2400 oilseed rape individuals collected along railway tracks throughout Switzerland proved positive for the presence of the transgenic Roundup Ready characteristic enzyme that confers glyphosate tolerance (Schoenenberger and D'Andrea 2012). Another study confirmed these findings and identified hot spots of transgenic plants at locations were unloading takes place (Hecht et al. 2014). These hot spots were ports and railway stations bordering to France and Italy, despite Switzerland's import ban (Schulze et al. 2015, 2014).

[5] www.transgen.de.

It can be considered as empirically confirmed that transgenes in oilseed rape are not limited to cultivation only, but can be found regularly as feral and as volunteer plants under a wide variety of geographic and ecological conditions.

Modelling Approaches for Gene Drives

There have been multiple attempts to model the gene flow of *B. napus*. Two of the most advanced approaches will be presented here. To model the population dynamics of a gene drive, possible modifications to existing models will be proposed. The two models explored are the GeneSys model established by Colbach et al. (2001a, b) and GeneTraMP by Middelhoff et al. (2011).

Another potentially relevant model was established by (Maxwell et al. 2015) which assumes a theoretical weed species and predicts the evolution and dynamics of herbicide resistance. On the basis of fungal spore dispersal, the model simulates pollen transfer between spatial entities assessing the interaction between herbicide application and plant genotype. Furthermore, Habekotté (1993, 1997a, b) established an empirical model BRASNAP-PH to predict key times of flowering and maturity of winter oilseed rape based on temperature and insolation. Some of her findings were included as input data in GeneSys and GeneTraMP.

GeneSys

The model was conceived to evaluate the effects of different cropping systems on gene flow in oilseed rape. It integrates the effects of crop succession and management at the level of a region. Simulation duration and crop succession and cultivation techniques are defined by the user. The model simulates the annual life cycle of cultivated and volunteer oilseed rape, starting with the seed bank at harvest and determining seedling emergence. Seedlings grow, flower and produce new seeds, some of which replenish the seed bank. At the end of each stage, the number of individuals per square meter is calculated. The seed bank is divided into four soil layers. Accordingly, seeds are distinguished by their respective layer and age, subdivided into young and old (older than one year) with related survival rates according to data by (Schlink 1994). The seeds' position in the soil layers may be changed by stubble breaking and the numbers of seeds and post-harvest seedlings may be reduced by pre-sowing tillage. Three tillage regimes are distinguished, stubble-breaking, ploughing and rigid-tine cultivation. Herbicide applications can be conducted after sowing in spring with mortality rates of 90 or 85%, respectively. Adult survival depends on vernalisation and density dependent intra- and interspecific competition. Newly formed seeds are deposited in the upper soil layer. The maximum adult oilseed rape plant density was taken from Freudhofmaier (1991) and Vullioud (1992), flower and seed production values from (Leterme 1985). The competitiveness of crop plants to volunteer plants was deduced by division of the maximum plant densities of the subsequent crop. Oilseed rape's seed loss rate can be specified as a variable parameter. Seedling emergence rates, effects of the sowing date, number of flowers and flower duration were experimentally assessed, considering tillage and intercrop periods. To integrate herbicide resistant varieties, the user can define the genotype as dominant or recessive, homozygous or heterozygous. Furthermore, self-pollination rate as well as seed and pollen production rate must be defined in accordance with the particular variety. The model deterministically considers pollen and seed transfer between fields and random import of foreign seed and contamination with foreign pollen. However, Couturaud (1998) shows that GeneSys tends to underestimate pollen and seed dispersal.

The model works with an aggregated plant representation and does not integrate individual plants, neither in nor outside the cropping area. Furthermore, it does not take differentiated climate response of the plants into consideration. Different genotypes in the model do not exhibit different developmental characteristics such as the onset of flowering. The model GeneSys has, however, been successfully applied in the definition of management rules for the commercial release of GM-oilseed rape crops in Europe (Bock et al. 2002; Messean et al. 2006).

GeneTraMP

The GeneTraMP model (Generic Transgene Movement and Persistence) was developed to assess regional effects of a large scale cultivation of GM-oilseed crops. Input data include results of landscape and climate analysis (Schmidt and Schröder 2011), satellite image processed oilseed rape distribution (Breckling et al. 2011) and regional crop management schemes (Glemnitz et al. 2011). This allows to analyse spatio-temporal dynamics of gene flow with implemented human crop management activities. The model differentiates between feral and volunteer plants as well as aggregated crop plant cohorts. Different genotypes, regarding cygosity, dominance or recessiveness and ploidy can be distinguished and implemented to simulate transgene spread. The model considers three types of pollen transfer: local pollen mass flow, insect-borne pollen transfer and unspecific regional background pollen transfer across larger distances. Pollination is dependent on the spatio-temporal distribution of flowering plants, the number of open flowers and their respective state of fertility. For the individual plant development, the model distinguishes five stages: seed, seedling, start of flowering, end of flowering and maturity. Progression through these stages is in accordance with climatic and environmental parameters. Seeds are further specified by one of four soil layers and their age, divided into young and old (older than one year), following (Nathalie Colbach et al. 2001). Also in accordance with GeneSys, seed persistence and germination decisions are calculated in dependence on temperature and moisture regimes. The temperature-dependent oilseed rape germination parameters were derived from (Habekotté 1997a) with the additional need for one rainy day in the week before germination. Plant development, flowering progress, flower and seed numbers were derived from Habekotté (1993, 1997a, b) and Cramer (1990).

Human activities are included in the form of cultivation practices, crop rotation, intensity and distribution of feral plants as a result of seed loss and road verge clearing activities. The model uses the same three tillage regimes as GeneSys. Weed control is implemented either as cutting, selective or total herbicide application. Information concerning crop management schedules and procedures in Germany were derived from a field cultivation survey (Zander 2003). The occurrence of volunteers is estimated for different subsequent crops in probability and density, according to field observations (Menzel 2006) and in dependence of environmental conditions. During oilseed rape harvest procedures, a stochastic amount of seeds is transferred to neighbouring grid cells, specified as ruderal areas. Feral oilseed rape plants are distinct from cultivation crop plants in frequency, size, number of inflorescences, flowering progress, pod forming rate and seed number per pod.

Model Suitability to Represent Gene Drive Population Dynamics

Both models are compared in Table 5.3. From these observations, it is obvious that the more recent GeneTraMP model, which in part is even based on the GeneSys model does not only contain more ecological details, but would be also more suitable to model the spread of gene drive oilseed rape plants. Even though, the model would have to be adjusted and requires further refinements for this purpose.

In order to use the GeneTramp model for the assessment of gene drive population dynamics and their spatio-temporal spread, certain additions to the program would have to be made. For instance, the model is already capable to simulate genotypes of transgenic varieties and their spread as volunteers and feral populations. For gene drive, the inheritance of such an organism as well as its genotypic characteristics would have to be implemented. Furthermore, potential side effects of the gene drive, such as an eventually reduced fitness would have to be implemented. At the current state of research, however, parameters concerning these side effects would be based on assumptions or require new empirical data. It would have to be decided whether a gene drive-carrier produces a mass of pollen or seeds that deviate from the conventional plants and whether and how gene drive organisms deviate in their life cycle. Apart from such considerations an even more significant issue might be to finalise the inclusion of oilseed rape's hybridisation partners which was not yet done due to the lack of appropriate data. Furthermore, if these were included, the next issue would be the consideration of introgression events. However, the GeneTraMP model was already successfully applied to simulate the occurrence of plants with multiple herbicide resistances emerging from single-resistant GM-oilseed rape crops cultivated in adjacent fields as was previously shown to occur in Canada (Hall et al. 2000). The model could be used to simulate not only a gene drive or the spread of resistance

Table 5.3 Comparison of the GeneSys and GeneTraMP models

Aspect	GeneSys	GeneTraMP
Regions	Small (Ile-de-France)	Large (Northern Germany), 4 different climate zones
Pollen transfer	Deterministic field to field, Random pollen import	Local pollen mass flow Insect borne transfer Regional long distance transfer
Seed transfer	Empirical values	–
Seed loss	User defined rate	Randomly determined from literature ranges
Plant development	Uniform from empirical data 4 seed stages, 2 seedling stages, 2–3 adult stages, flowers, 2 new seed stages	Temperature and precipitation dependent seed, seedling, start of flowering, end of flowering, maturity
Crop rotation	User defined	10 most widely applied generic, area specific rotations
Cultivation densities	–	3
Genotypes	Dominant-recessive, Cygosity	Dominant-recessive, Cygosity, Ploidy
Herbicides	Up to two (unspecified)	One (unspecified)
Crop management	User defined, Tillage, cutting, chisel, herbicide treatment	Included Ploughing, rigid tine, both
Populations	Crop, volunteer	Crop, volunteer, feral
Soil layers	4	4, analogous to GeneSys
Hybridisation partners	–	Not yet fully included

genes but also the combination of both and the ultimate formation of resistance-carrying gene drive-carriers. Given the necessary input, it would even be possible to model the application of gene drives to target and remove herbicide resistances in volunteer and feral populations.

Suitability and Prerequisites of Oilseed Rape for Gene Drives?

Ideally, the target organisms for a gene drive would exist in geographically isolated populations. Wherein, the migration of individuals from one population to another does (at best) not occur. Target organisms must reproduce sexually with relatively short generation times. Non-overlapping generations would thereby allow an easier assessment of the gene drive's spread in the population. Furthermore, the target organism should be diploid and its genome should already be well studied, in the context of gene expression patterns over the course of its development, while sequence variations within the targeted genes should be rare. The mutation rate of the target organism should also be low. Moreover, there should be no interspecific mating events, barring the creation of gene drive-carrying hybrids. Lastly, the application of a gene drive is a risky intervention into, not only the genome of a target species but also its ecosystem, which may harbour unforeseen consequences. Therefore, experts recommend to only consider (at least CRISPR/Cas-based gene drives) for applications that, if successful, have an extraordinarily high level of benefit for large majorities. Such aims include the population control of vector species for diseases, such as for malaria. Multiple approaches to tackle the disease have been suggested (Burt et al. 2018). The application of gene drives for conservational goals such as invasive species control on the other hand, is controversial and considered too risky—or for some experts the risks would outweigh the intended benefits (Esvelt and Gemmell 2017).

According to several already mentioned aspects, it is foreseeable that oilseed rape does not fit well the criteria for an ideal gene drive target organism. This is further illustrated in Table 5.4. Oilseed rape is an allotetraploid plant complicating the engineering and propagation of a potent gene drive system, albeit the genome is well studied also due to the many available transgenic variants. The genetic variability of modern oilseed rape varieties are low due to the two bottlenecks in selective breeding for low erucic acid and low glucosinolate content (Bus et al. 2011). However, in a study of the allelic diversity of 72 varieties of oilseed rape from five different countries, Chen et al. (2007) found a total of 59 private alleles of which 21 are shared with other *Brassica* species. On the one hand, the spread of gene drive pollen would be partially reduced due to the fact that oilseed rape is self-compatible. On the other hand, spatial confinement of a gene drive would be nearly impossible as oilseed rape is cultivated around the world and many feral populations exist, mainly due to seed losses along transportation routes. Additionally, pollen is dispersed over long distances. The occurrence of feral populations also makes the confinement of

Table 5.4 Requirements for an ideal gene drive target organism

Ideal target organism	Oilseed rape
Diploid	Allotetraploid
Geographically isolated	Cultivated around the world
Low genetic variability	Low due to selective breeding
Low migration rate	Long range pollen dispersal, transportation losses
Low mutation rate and genetic diversity	?
No interspecific gene flow	Interspecific gene flow with introgression
Non-overlapping generations	Overlapping generations due to perennial Feral plants and secondary dormancy
Sexual reproduction	High self-compatibility
Short generation times	Annual or even perennial
Well studied genome	Many transgenic variants

a gene drive to the target population and even target species difficult, as the plant possesses many hybridisation partners. Temporal confinement is a major problem due to the long generation times and even overlapping generations in the case of perennial stocks. This last point is even further aggravated due to the secondary dormancy, which allows certain seeds to remain dormant in seed banks for over a decade (Pekrun 1994).

An even greater problem is that oilseed rape is an agricultural product. It is therefore likely that gene drive pollen could fertilize conventionally grown oilseed rape. On the one hand, this would reduce the spread of the gene drive as these plants are harvested annually. But on the other hand it likely will give rise to gene drive volunteer plants which may be detrimental to organic farmers and socially not well-received even in conventional cropping.

Purposes for Oilseed Rape Gene Drives?

In general, gene drives can be subdivided by their intended goal to either suppress or replace/modify wild populations, to disperse certain alleles in a population, or to limit genetic exchange with other populations. In most techniques, this is decided by the cargo genes associated with the drive mechanism. Suppression might, for instance, be achieved by driving stimulus specific vulnerabilities (to either low or high temperatures or certain chemicals) into the population. A modification to suppress volunteering without suppressing the population might entail the deletion or repression of the genes responsible for secondary dormancy. Suppression could also be achieved by infertility. Especially female infertility may be a desirable trait to prevent the hybridisation with other plant species, as hybridisation with oilseed rape is mostly successful with *B. napus* as the female parent (Scheffler and Dale 1994). Organisms with infertile female sex organs would also not produce seeds but still produce and disperse pollen and could thereby propagate the trait.

As already predicted in 2004 and even before, the cultivation of transgenic oilseed rape has led to transgene escape into feral populations (Breckling and Menzel 2004). As the examples of five different countries and the EU show, transgenic contamination and concomitant dispersal of herbicide resistance genes is a serious issue which supports an invasion of feral transgenic oilseed rape into many ecosystems. This eventually leads to the potential gene flow and introgression of transgenic herbicide resistances to other hybridisation partners of oilseed rape. This may negatively affect the integrity of genetic resources in wild relatives (Londo et al. 2010).

Furthermore, we are currently experiencing an anthropomorphic downward spiral in agriculture, called the pesticide treadmill. Due to the broad application of herbicides weeds with single or multiple resistances, so called superweeds emerge. Therefore, the development of new generations of pesticides is required for maintaining of pesticide protected industrial agriculture. These will however, eventually be followed by yet another generation of superweeds and the vicious cycle continues [For a scope of the pesticide treadmills governance issues see Bain et al. (2017)].

Gene drive applications might potentially stop or at least slow down this vicious cycle, but may also cause new problems.

This section is therefore dedicated to the exploration of conceivable gene drive applications with the goal to get rid of transgenic feral and volunteer populations of oilseed rape.

Gene Drives to Delete/Block Herbicide Resistances

We consider the possibility to eliminate or reduce the occurrence of herbicide resistance from undesirable escape of transgenes to the wild and their persistence in feral populations. To reduce the occurrence of herbicide-resistant feral populations, an obvious target locus for a gene drive would be the resistance genes themselves. Despite heavy outcrossing of the feral populations, the resistance genes confer a fitness gain and are therefore likely to become conserved.

CRISPR/Cas-Based Approaches

It might seem at least feasible to use a CRISPR/Cas-based gene drive to target all different herbicide resistance-loci in order to delete or disrupt them by homing cargo genes into them. The gene drive would confer a fitness penalty, which might cause the gene drive carrying oilseed rape to vanish after a number of generations. Multiple releases over several years might be necessary. Furthermore, this fitness penalty would favour the selection for gene drive resistant alleles. However, considering the large number of herbicide resistant variants this would require many gRNAs, most likely resulting in a higher genetic load which may confer an even higher fitness penalty.

The formation of resistance alleles, due to non-homologous end joining instead of homology-directed repair, would on one hand impede the copying of the gene drive cassette, but may on the other hand cause a null-mutation in the present resistance allele. This would constitute a positive side effect of imperfect repair not present in other gene drive approaches currently discussed.

Probably, the most problematic issue of this approach is the potential cross-fertilisation of (feral) gene drive oilseed rape with cultivated transgenic varieties. The seeds of these crosses would also carry the gene drive and might remain in the seed bank for decades being able to restart the gene drive. Furthermore, this gene drive oilseed rape may occur as volunteer in subsequent crop rotations. Another problematic aspect would be the risk of an outcrossing to related species.

Resistance Blocking

An alternative pathway could include either miRNA cargo genes that target the resistance genes' mRNAs or cargo genes for proteins that either repress the gene expression of the resistances or inactivate the gene product post-translationally. While a CRISPR/Cas-based approach would act only once in the germline, this alternative approach would require a constitutive gene expression throughout the life cycle of the plant to block resistance. These cargo genes would therefore confer a high fitness penalty and thus be subject to mutational changes. Since a single base pair deletion could lead to reading frame shifts in the genes of proteins, this may be detrimental to the gene drive and may in the end result in carrier-organisms retaining partial resistance. This effect would probably be less pronounced when using miRNAs to block resistance instead of proteins. But then, multiple RNAs should be included within the gene drive cassette to silence a single resistance gene. But despite a successful gene drive application, the population would then still retain the resistance genes.

Overwriting Various Herbicide Resistances with a Single Resistance

It might be feasible to equip the gene drive with resistance against a newly designed herbicide including the implication to overwrite or disrupt existing resistances. If such a gene drive would be used in cultivation and successively spread to feral populations, it might be an approach to eliminate herbicide resistances from feral populations. This, however, would be a long-term and large scale intervention.

It is obvious that such an approach is based on some major assumptions which seem difficult to achieve. Another problematic issue would be the almost impossible monitoring of such a gene drive's spread in feral populations. Moreover, it would be necessary to sow gene drive oilseed rape into identified feral populations which is especially time-consuming and makes the applications of a gene drive hardly useful, as the feral population could simply be destroyed instead.

Self-Limiting CRISPR/Cas-Based Approaches

A more confineable CRISPR-based technique would simply cut out the resistance alleles without propagating the CRISPR/Cas-system itself. This self-limiting approach would, therefore, not constitute a gene drive. Thus, multiple mass-releases would be necessary, although their seeds may remain in the seed bank for multiple generations as well. However, these individuals carrying non-functional resistance genes would be less fit than their resistant conspecifics and probably even less fit than conventional non-GMO oilseed rape stocks.

Finally, there is the possibility of self-limiting CRISPR/Cas-based gene drives, e.g. the daisy-chain drive. A daisy drive system consists of multiple gene drives in which each element drives the next element in the chain. The bottom element is lost from the population first, then the next element ceases to drive and gets lost from the population and so on. This process continues upwards in the chain until, eventually, the population returns to its wild-type state with regard to the expression of genes expressing intended artificial functionalities (Noble et al. 2016). This approach would limit the spread of a gene drive in time and space which on one hand would increase controllability, but on the other hand would be less desired in the case of the widely spread feral populations.

Gene Drive Seeds as a Breeding Tool on Agricultural Fields

To take this thought one step further, though ecologically not really desirable, it might be feasible to sow gene drive carrying oilseed rape on agricultural fields. Although it seems unlikely to have farmers being willing to sow gene drive-carrying oilseed rape within the EU, there may be functionalities of such an approach. The characteristics of the gene drive could be that of the resistance-overwriting approach which would overwrite existing herbicide resistances (and gene drives) with a single resistance. This might also help in accelerating breeding procedures towards the reproduction of basic seed. After a number of generations, taking into account the spread of the gene drive to feral populations and the persistence in the seed banks, a switch of overwriting drive carrying oilseed rape and herbicide might be performed. The advantage would be the seed losses during transportation, which would guarantee a wide spread of the gene drive and enable it to reach more distant feral populations.

While currently, new herbicides have to be developed again and again to overcome the emerging formation and outcrossing of resistances, this approach would allow to alternate between a limited set of herbicides, if extended towards additional weed species. Since the massive use of a limited number of active herbicide ingredients led to the selection of major problematic weeds worldwide and in particular in those agricultural systems using herbicides extensively (Heap 2014). The alternative in organic agriculture would be adequate crop rotations and the sustaining of a rich biodiversity in the surrounding landscape to limit pest outbreaks.

Female Infertility Drive

This paragraph serves to further explore the thought experiments of a gene drive conferring female infertility. As mentioned above, this could be achieved by repressing or even deleting the genes responsible for the development and functioning of the plants' stigma or gamete production. To be able to mass-produce these organisms in the laboratory, it needs to be equipped with a repressor mechanism, which would allow normal development in the presence of a certain chemical, as is the case in the Release of Insects carrying a Dominant Lethal (RIDL)-system (Thomas et al. 2000). Assuming again, a drive propagation by the transformative action of a CRISPR/Cas-based gene drive, a high conversion rate could be expected. The drive would only propagate via pollen. It would not remain in the seed bank, because it is unable to produce seeds. In order to pursue the set goal of the suppression of transgenic feral populations, the drive could again home in on various resistance gene sequences. Thus, only the offspring of transgenic parents would be converted to homozygous carriers, while non-GMO parents would give rise to heterozygous offspring. The latter would be the case, if the gene drive pollen was to fertilize cultivated plants. Assuming for the female infertility to only homozygously be effective, the cross-pollination of cultivated gene drive crops would not pose a threat to conventional agriculture. Instead, it would reduce the occurrence of volunteer plants in subsequent crop rotations. Furthermore, this would cause the drive to be almost self-limiting, as it would stop its propagation, if no transgenic plants are fertilized for several generations. However, in this scenario, multiple releases over several decades would again be necessary.

Other Gene Drive Techniques Besides CRISPR/Cas

Up to this point, the considered gene drive applications revolved around drives propagated by the CRISPR/Cas system. To complete the available drive systems, we discuss them below.

Medea (Maternal Effect Dominant Embryonic Arrest)

This technique uses a toxin-antidote combination that causes it to be inherited with a super-Mendelian frequency. The maternally expressed toxin is deposited in the embryo and only if that embryo inherits the gene drive, it will be able to counteract the toxin with the appropriate antidote. This functionality was discovered occurring naturally in *Tribolium* flour beetles. It was first synthetically engineered in *Drosophila* (Akbari et al. 2014). In principle, Medea constitutes a modification drive designed to be able to replace wild populations.

Underdominance

The phenomenon of Underdominance causes heterozygous carriers to receive a higher fitness penalty than homozygous carriers. This can be achieved by toxin-antidote combinations, either similar to the Medea technique (Akbari et al. 2013), or containing an miRNA toxin, targeting a vital gene, and a recoded but haplo-insufficient version of the gene (Reeves and Reed 2015). Underdominance constitutes a modification drive as well, but different from Medea depends on a higher threshold within a population in order to drive the desired traits to fixation.

Killer-Rescue

Independently inherited toxin (killer) and antidote (rescue) alleles will transiently increase the rescue alleles (fused to cargo genes) prevalence within a population. The lethal killer alleles will be quickly selected from the population and as soon as this happens, the rescue alleles will no longer confer a fitness gain and therefore, will be reduced in frequency in the population as well (Gould et al. 2008).

All three toxin-antidote techniques may potentially be applicable for gene drives in oilseed rape. Albeit up until now, none of these techniques has been developed for plant species. Whether their functionality in allotetraploid plants would be comparable to that in diploid animals is therefore unknown. The ideas presented above, could be incorporated into these drive mechanisms as well with the exception of the female infertility trait in combination with the Medea or the Medea-related Underdominance drive systems because these techniques rely on a maternally deposited toxin to select against offspring not carrying the drive construct(s).

Meiotic drives

These naturally occurring drive systems rely on the distortion of chromosome segregation during meiosis. The most widely discussed approach in this category is the X-Shredder technique, which in sexually reproducing animals distorts the sex ratio by targeting the X-chromosome with nucleases. If the necessary genes are located on the Y-chromosome, this can result in a male bias of up to 95% (Galizi et al. 2014). Although the sex determination in plants is more complex than in animals and oilseed rape is monoecious, as each plant has both sexes united in perfect flowers, the Bassicaceae family belongs to the group of eurosids II, which were shown to possess sex chromosomes (Charlesworth 2002). Therefore, it might be feasible to engineer a drive system resembling the X-Shredder technique in oilseed rape. Wherein, carrier-organisms may produce a high percentage of androcious offspring. These would be able to propagate via pollen.

This drive, however, useful to suppress feral populations would again be problematic due to the fertilisation of cultivated crops. The consequence would be gene drive-carrying volunteers. Furthermore, by hybridisation events, this drive might invade other species, distorting their sex ratios as well. This would most likely prove detrimental for the flora with potential secondary effects to the fauna of a vast number of ecosystems. However unlikely, such an event seems probable considering the long lasting persistence of the drive due to volunteering, formation of feral populations and long-term persistence in the seed bank inherent to oilseed rape. Even events with an extremely low probability become significant, given enough time and a large sample size (Diaconis and Mosteller 1989).

Conclusion

This case study presented oilseed rape in its agricultural significance, its ecological interconnections as well as its survival and propagation strategies. It pointed to the numerous potential hybridisation partners, lists some of the transgenic varieties and stressed the emerging problem of escaping transgenes into the environment. Existing computer programs applicable to model the gene flow of oilseed rape and the spread of those escaping transgenes were reviewed and compared, especially points of improvement in the light of the simulation of gene drive were emphasised. Finally, the suitability of oilseed rape as a target organism for gene drives was scrutinized and furthermore multiple ideas for the various, currently discussed gene drive approaches were exercised in thought experiments in various modes with different potential cargo genes.

In summary, the application of a gene drive to oilseed rape as target organism would be laborious to engineer, very difficult to confine, almost impossible to adequately monitor and would require multiple mass releases across large areas and long time spans. Such an endeavour would be very costly. The risk of losing control would increase over time, and considering the long time frame would almost be certain. In face of the (apparently mild or negligible) benefits, a successful gene drive would harbour the reduction of transgene escape into feral populations or, economically more interesting, potentially slowing down the pesticide treadmill. These circumstances make the application of an oilseed rape gene drive an unfavourable bargain. Therefore, such an application is not recommended. It was demonstrated that various risk dimensions are required to be considered. Among these are stability and dispersal issues, gene flow and complex spatial crop-volunteer-feral plant interactions, not to mention conservation and public acceptance issues. Although oilseed rape is not a good target organism for gene drive applications, it points out a panoply of relevant issues that may arise in part or in various combinations, when considering other possible plant target organisms for gene drive.

The major value of the considerations is not the expectation of a near-future application of SPAGE in oilseed rape. Instead, we intended this case study to emphasize the diverse issues which are crucial for an implication analysis in plants. Oilseed rape is an excellent example for such an exercise, since it allows to exemplify the dimensions of the physiological, ecological, environmental and agricultural context, which is relevant also for the risk assessment of other potential botanical candidates for genetic modification of the considered types.

References

Adler, L. S., Wikler, K., Wyndham, F. S., Linder, C. R., & Schmitt, J. (1993). Potential for persistence of genes escaped from canola: germination cues in crop, wild, and crop-wild hybrid *Brassica rapa*. *Functional Ecology*, 736–745.

Adolphi, K. (1995). *Neophytische Kultur-und Anbaupflanzen als Kulturflüchtlinge des Rheinlandes*. Galunder.

Akbari, O. S., Chen, C.-H., Marshall, J. M., Huang, H., Antoshechkin, I., & Hay, B. A. (2014). Novel synthetic Medea selfish genetic elements drive population replacement in drosophila, and a theoretical exploration of Medea-dependent population suppression. *ACS Synthetic Biology, 3*, 015–928.

Akbari, O. S., Matzen, K. D., Marshall, J. M., Huang, H., Ward, C. M., & Hay, B. A. (2013). A synthetic gene drive system for local, reversible modification and suppression of insect populations. *Current Biology, 23*, 671–677. https://doi.org/10.1016/j.cub.2013.02.059.

Ammann, K., & Vogel, B. (1999). Langzeitmonitoring gentechnisch veränderter Organismen. Bestandsaufnahme, Fallbeispiele und Empfehlungen. Kantonales Laboratorium Basel-Stadt Kontrollstelle für Chemie und Biosicherheit (KCB) 70.

Aono, M., Wakiyama, S., Nagatsu, M., Kaneko, Y., Nishizawa, T., Nakajima, N., et al. (2011). Seeds of a possible natural hybrid between herbicide-resistant *Brassica napus* and *Brassica rapa* detected on a riverbank in Japan. *GM Crops, 2*, 201–210.

Aono, M., Wakiyama, S., Nagatsu, M., Nakajima, N., Tamaoki, M., Kubo, A., et al. (2006). Detection of feral transgenic oilseed rape with multiple-herbicide resistance in Japan. *Environmental Biosafety Research, 5*, 77–87.

Augustin, C., Becker, R., Gottwald, R., Hedtke, C., Hornermeier, B., Lentzsch, P., et al. (1998). Ökologische Auswirkungen der Einführung der Herbizidresistenz (HR)-Technik bei Raps und Mais. - Gutachten für das Ministerium für Umwelt, Naturschutz und Raumordnung (MUNR) des Landes Brandenburg. Eigenverlag.

Bain, C., Selfa, T., Dandachi, T., & Velardi, S. (2017). 'Superweeds' or 'survivors'? Framing the problem of glyphosate resistant weeds and genetically engineered crops. *Journal of Rural Studies, 51*, 211–221.

Baranger, A., Chevre, A. M., Eber, F., & Renard, M. (1995). Effect of oilseed rape genotype on the spontaneous hybridization rate with a weedy species: An assessment of transgene dispersal. *Theoretical and Applied Genetics, 91*, 956–963.

Bauer-Panskus, A., Hamberger, S., Then, C. (2013). Transgene Escape–Atlas der unkontrollierten Verbreitung gentechnisch veränderter Pflanzen.

Becker, T. H. (1951). Siebenjährige blütenbiologische Studien an den Cruziferen *Brassica napus* L., *Brassica rapa* L., *Brassica oleracea* L., *Raphanus* L. und *Sinapis* L.. *Teil I. Zeitschrift für Pflanzenzüchtung, 29*, 222–240.

Beckie, H. J., & Warwick, S. I. (2010). Persistence of an oilseed rape transgene in the environment. *Crop Protection, 29*, 509–512.

Bijral, J. S., & Sharma, T. R. (1996). Intergeneric hybridization between *Brassica napus* and *Diplotaxis muralis*. *Cruciferae Newslett. Eucarpia, 18*, 10–11.

Bing, D. J., Downey, R. K., & Rakow, G. F. W. (1991). Potential of gene transfer among oilseed Brassica and their weedy relatives. *GCIRC Rapeseed Congr., 1991*, 1022–1027.

Bock, A.-K., Lheureux, K., Libeau-Dulos, M., Nilsagard, H., & Rodriguez-Cerezo, E. (Eds.), (2002). *Scenarios for co-existence of genetically modified, conventional and organic crops in European agriculture*. IPTS-JRC study.

Braatz, J., Harloff, H.-J., Mascher, M., Stein, N., Himmelbach, A., & Jung, C. (2017). CRISPR-Cas9 targeted mutagenesis leads to simultaneous modification of different homoeologous gene copies in polyploid oilseed rape (*Brassica napus*). *Plant Physiology, 174*, 935. https://doi.org/10.1104/pp.17.00426.

Breckling, B., Borgmann, P., Neuffer, B., Hurka, H., Tappeser, B., Brauner, R., Menzel, G., Born, A., Schmidt, G., Schröder, W., Laue, H., Middelhoff, U., Baumann, R., Keßler, M., Reiche, E.-W.,

Rinker, A., Tillmann, J., Windhorst, W., Glemnitz, M., Wurbs, A., Funke, B., & Brozio, S. (2004). GenEERA - Forschungsverbund Generische Erfassung und Extrapolation der Raps-Ausbreitung (Abschlussbericht). Universität Kiel.

Breckling, B., Laue, H., & Pehlke, H. (2011). Remote sensing as a data source to analyse regional implications of genetically modified plants in agriculture-oilseed rape (*Brassica napus*) in Northern Germany. *Ecological Indicators, 11*, 942–950.

Breckling, B., & Menzel, G. (2004). Self-organised pattern in oilseed rape distribution—An issue to be considered in risk analysis. In B. Breckling & R. Verhoeven (Eds.), *Risk hazard damage, specification of criteria to assess environmental impact of genetically modified organisms* (pp. 73–88). Bundesamt für Naturschutz, Bonn.

Brown, J., & Brown, A. P. (1996). Gene transfer between canola (*Brassica napus* L. and *B. campestris* L.) and related weed species. *Annals of Applied Biology, 129*, 513–522.

Burt, A., Coulibaly, M., Crisanti, A., Diabate, A., & Kayondo, J. K. (2018). Gene drive to reduce malaria transmission in Sub-Saharan Africa. *Journal of Responsible Innovation, 5*, S66–S80. https://doi.org/10.1080/23299460.2017.1419410.

Bus, A., Körber, N., Snowdon, R. J., & Stich, B. (2011). Patterns of molecular variation in a species-wide germplasm set of *Brassica napus*. *Theoretical and Applied Genetics, 123*, 1413–1423. https://doi.org/10.1007/s00122-011-1676-7.

Chalhoub, B., Denoeud, F., Liu, S., Parkin, I. A., Tang, H., Wang, X., et al. (2014). Early allopolyploid evolution in the post-Neolithic *Brassica napus* oilseed genome. *Science, 345*, 950–953.

Charlesworth, D. (2002). Plant sex determination and sex chromosomes. *Heredity, 88*.

Chen, B. Y., Heneen, W. K., & Jönsson, R. (1988). Resynthesis of *Brassies napus* L. through inter-specific hybridization between *B. alboglabra* Bailey and *B. campestris* L. with special emphasis on seed colour. *Plant Breeding, 101*, 52–59.

Chen, S., Nelson, M. N., Ghamkhar, T., Fu, T., & Cowling, W. A. (2007). Divergent patterns of allelic diversity from similar origins: the case of oilseed rape (*Brassica napus L.*) in China and Australia. *Genome, 51*, 1–10. https://doi.org/10.1139/G07-095.

Chèvre, A. M., Ammitzbollz, H., Breckling, B., Dietz-Pfeilstetter, A., Eber, F., Meier, M. S., Den NiJs, H. C. M., Pascher, K., Seguin-Swartz, G., Sweet, J. B., Stewart, C. N. J., & Warwick, S. I. (2004). A review on interspecific gene flow from oilseed rape to wild relatives. In *Introgression from genetically modified plants into wild relatives*. CABI Publishing, Cambridge.

Chèvre, A.-M., Eber, F., Baranger, A., & Renard, M. (1997). Gene flow from transgenic crops. *Nature, 389*, 924.

Colbach, N., Clermont-Dauphin, C., & Jean-Marc, M. (2001). *GENESYS: A model of the influence of cropping system on gene escape from herbicide tolerant rapeseed crops to rape volunteers—I. Temporal evolution of a population of rapeseed volunteers in a field.*

Colbach, N., Clermont-Dauphin, C., & Meynard, J. M. (2001). GeneSys: A model of the influence of cropping system on gene escape from herbicide tolerant rapeseed crops to rape volunteers: II. Genetic exchanges among volunteer and cropped populations in a small region. *Agriculture, Ecosystems & Environment, 83*, 255–270. https://doi.org/10.1016/S0167-8809(00)00175-4.

Conservation Council of Western Australia CCWA. (2012). *A survey of roadside fugitive GM (Roundup Ready) canola plants at Williams*. CCWA: Western Australia.

Couturaud, M. C. (1998). *Effet des systèmes de culture sur les risques de dissémination du transgène de colza dans l'environnement: évaluation et utilisation du modèle GENESYS*. INA PG, Paris: Mémoire de DAA.

Cramer, N. (1990). Raps: Züchtung, Anbau und Vermarktung von Körnerraps. Stuttgart, Ulmer. S 146.

Crawley, M. J., & Brown, S. L. (2004). Spatially structured population dynamics in feral oilseed rape. *Proceedings of the Royal Society of London B: Biological Sciences, 271*, 1909–1916.

Crawley, M. J., Hails, R. S., Rees, M., Kohn, D., & Buxton, J. (1993). Ecology of transgenic oilseed rape in natural habitats. *Nature, 363*, 620–623.

Darmency, H., Lefol, E., & Fleury, A. (1998). Spontaneous hybridizations between oilseed rape and wild radish. *Molecular Ecology, 7*, 1467–1473.

Devaux, C., Lavigne, C., Falentin-Guyomarc'h, H., Vautrin, S., Lecomte, J., & Klein, E. K. (2005). High diversity of oilseed rape pollen clouds over an agro-ecosystem indicates long-distance dispersal. *Molecular Ecology, 14*, 2269–2280.

Devos, Y., De Schrijver, A., & Reheul, D. (2009). Quantifying the introgressive hybridisation propensity between transgenic oilseed rape and its wild/weedy relatives. *Environmental Monitoring and Assessment, 149*, 303–322.

D'Hertefeldt, T., Jørgensen, R. B., & Pettersson, L. B. (2008). Long-term persistence of GM oilseed rape in the seedbank. *Biology Letters, 4*, 314–317.

Diaconis, P., & Mosteller, F. (1989). Methods for studying coincidences. *Journal of the American Statistical Association, 84*, 853–861.

Eber, F., Chèvre, A. M., Baranger, A., Vallée, P., Tanguy, X., & Renard, M. (1994). Spontaneous hybridization between a male-sterile oilseed rape and two weeds. *Theoretical and Applied Genetics, 88*, 362–368.

Eckert, J. E. (1933). The flight range of the honeybee. *Journal of Agricultural Research.*

Eschmann-Grupe, G., Hurka, H., & Neuffer, B. (2003). Species relationships within Diplotaxis (Brassicaceae) and the phylogenetic origin of *D. muralis*. *Plant Systematics and Evolution, 243*, 13–29.

Esvelt, K. M., & Gemmell, N. J. (2017). Conservation demands safe gene drive. *PLOS Biology, 15*, e2003850.

Feil, B., & Schmid, J. E. (2001). Pollenflug bei Mais, Weizen und Roggen. Ein Beitrag zur Frage der beim Anbau von transgenen Kulturpflanzen erforderlichen Sicherheitsabstände. Institut f. Pflanzenbauwissenschaften, ETH Zürich.

Fischbeck, G. (1998). Sicherheitsforschung zu Freisetzungsversuchen in Roggenstein. Einführung und Ergebnisse zur Pollen- und Samenverbreitung transgener Erbeigenschaften. Verband Deutscher Biologen: Gentechnik, Ökologie und Ernährung, München 5–8.

Fischbeck, G., Heyland, K.-U., & Knauer, N. (1982). *Spezieller Pflanzenbau* (2nd ed.). Stuttgart: Eugen Ulmer.

Förster, K., Schuster, C., Belter, A., & Diepenbrock, W. (1998). Agrarökologische Auswirkungen des Anbaus von transgenem herbizidtoleranten Raps (*Brassica napus* L.). *Bundesgesundheitsblatt, 41*, 547.

Franzaring, J., Holz, I., Fangmeier, A., & Zipperle, J. (2008). Monitoring the absence of glyphosate and glufosinate resistance traits in feral oilseed rape and wild crucifer populations. In Breckling, B., Reuter, H., & Verhoeven, R. (Eds.), *Implications of GM crop cultivation at large spatial scales* (Vol. 14, pp. 90–92). Theorie in der Ökologie.

Freudhofmaier, O. (1991). *Standraum und Ertragsstruktur von Raps. Raps, 9*, 148–152.

Friesen, L. F., Nelson, A. G., & Van Acker, R. C. (2003). Evidence of contamination of pedigreed canola (*Brassica napus*) seedlots in western Canada with genetically engineered herbicide resistance traits. *Agronomy Journal, 95*, 1342–1347.

Galizi, R., Doyle, L.A., Menichelli, M., Bernardini, F., Deredec, A., Burt, A., Windbichler, N., & Crisanti, A. (2014). A synthetic sex ratio distortion system for the control of the human Malaria mosquito. *Nature Communications, 5*, 3977. https://doi.org/1038/ncomms4977

Garnier, A., Pivard, S., & Lecomte, J. (2008). Measuring and modelling anthropogenic secondary seed dispersal along roadverges for feral oilseed rape. *Basic and Applied Ecology, 9*, 533–541.

Gerdemann-Knörck, M., & Tegeder, M. (1997). Kompendium der für Freisetzungen relevanten Pflanzen; hier: Brassicaceae, Beta vulgaris, Linum usitatissimum. Texte, 38/97, Umweltbundesamt Berlin 221pp.

Gland, A. (1982). Gehalt und Muster der Glucosinolate in Samen von resynthetisierten Rapsformen. *Z. Pflanzenzüchtung, 87*, 613–617.

Glemnitz, M., Wurbs, A., & Roth, R. (2011). Derivation of regional crop sequences as an indicator for potential GMO dispersal on large spatial scales. *Ecological Indicators, 11*, 964–973.

Gliddon, C. J. (1999). *Gene flow and risk assessment.*

Gould, F., Huang, Y., Legros, M., & Lloyd, A. L. (2008). A Killer-Rescue system for self-limiting gene drive of anti-pathogen constructs. *Proceedings of the Royal Society B: Biological Sciences, 275*, 2823–2829. https://doi.org/10.1098/rspb.2008.0846.

Gressel, J. (1999). Tandem constructs: Preventing the rise of superweeds. *Trends in Biotechnology, 17*, 379–388.

Habekotté, B. (1997a). Evaluation of seed yield determining factors of winter oilseed rape (*Brassica napus* L.) by means of crop growth modelling. *Field Crops Research, 54*, 137–151.

Habekotté, B. (1997b). A model of the phenological development of winter oilseed rape (*Brassica napus* L.). *Field Crops Research, 54*, 127–136.

Habekotté, B. (1993). Quantitative analysis of pod formation, seed set and seed filling in winter oilseed rape (*Brassica napus* L.) under field conditions. *Field Crops Research, 35*, 21–33.

Halfhill, M. D., Millwood, R. J., Raymer, P. L., & Stewart, C. N., Jr. (2002). Bt-transgenic oilseed rape hybridization with its weedy relative, *Brassica rapa. Environmental Biosafety Research, 1*, 19–28.

Hall, L., Topinka, K., Huffman, J., Davis, L., & Good, A. (2000). Pollen flow between herbicide-resistant *Brassica napus* is the cause of multiple-resistant *B. napus* volunteers. *Weed Science, 48*, 688–694.

Hansen, L. B., Siegismund, H. R., & Jørgensen, R. B. (2001). Introgression between oilseed rape (*Brassica napus* L.) and its weedy relative *B. rapa* L. in a natural population. *Genetic Resources and Crop Evolution, 48*, 621–627.

Harberd, D. J., & McArthur, E. D. (1980). *Meiotic analysis of some species and genus hybrids in the Brassiceae. Meiotic analysis of some species and genus hybrids in the Brassiceae.*

Heap, I. (2014). Herbicide resistant weeds. In D. Pimentel, R. Peshin (Eds.), *Integrated pest management* (pp. 281–301). Springer Netherlands, Dordrecht. https://doi.org/10.1007/978-94-007-7796-5_12

Hecht, M., Oehen, B., Schulze, J., Brodmann, P., & Bagutti, C. (2014). Detection of feral GT73 transgenic oilseed rape (*Brassica napus*) along railway lines on entry routes to oilseed factories in Switzerland. *Environmental Science and Pollution Research, 21*, 1455–1465.

Hofmann, N., Neubert-Kester, G., Gellermann, M., & Thienel, T. (2007). Untersuchungen zur Verbreitung und Anreicherung von Transgensequenzen in der Umwelt über Auskreuzung und Bodeneintrag am Beispiel von HR-Raps. Bundesamt für Naturschutz.

Hühn, M., & Rakow, G. (1979). Einige experimentelle Ergebnisse zur Fremdbefruchtungsrate bei Winterraps (*Brassica napus* oleifera) in Abhangigkeit von Sorte und Abstand. Zeitschrift fur Pflanzenzuchtung.= Journal of plant breeding.

Hurka, H., Bleeker, W., & Neuffer, B. (2003). Evolutionary processes associated with biological invasions in the Brassicaceae. *Biological Invasions, 5*, 281–292.

Ingram, J. (2000). The separation distances required to ensure cross-pollination is below specified limits in non-seed crops of sugar beet, maize and oilseed rape. *Plant Varieties & Seeds, 13*, 181–199.

Klinger, T., & Ellstrand, N. C. (1994). Engineered genes in wild populations: Fitness of weed-crop hybrids of *Raphanus Sativus. Ecological Applications, 4*, 117–120.

James, C. (2015). *Global status of commercialized biotech/GM crops: 2014*. ISAAA brief 49.

Janchen, E. (1972). *Flora von Wien* (p. 353). Verein für Landeskunde von Niederösterreich und Wien: Niederösterreich und Nordburgenland.

Jørgensen, R. B. (1999). *Gene flow from oilseed rape (Brassica napus) to related species.*

Jørgensen, R. B., & Andersen, B. (1994). Spontaneous hybridization between oilseed rape (*Brassica napus*) and weedy *B. campestris* (Brassicaceae): A risk of growing genetically modified oilseed rape. *American Journal of Botany*, 1620–1626.

Jørgensen, R. B., Andersen, B., Landbo, L., & Mikkelsen, T. R. (1996.) Spontaneous hybridization between oilseed rape (*Brassica napus*) and weedy relatives. In *Acta Horticulturae. International Society for Horticultural Science (ISHS)* (pp. 193–200), Leuven, Belgium. https://doi.org/10.17660/ActaHortic.1996.407.23

Jørgensen, R. B., Hauser, T., Mikkelsen, T. R., & Østergård, H. (1996). Transfer of engineered genes from crop to wild plants. *Trends in Plant Science, 1*, 356–358.

Karpechenko, G. D. (1924). Hybrids of ♀*Raphanus sativus* L. x ♂*Brassica oleracea* L. *Journal of Genetics, 14*, 375–396.

Kawata, M., Murakami, K., & Ishikawa, T. (2009). Dispersal and persistence of genetically modified oilseed rape around Japanese harbors. *Environmental Science and Pollution Research, 16*, 120–126.

Kerlan, M. C., Chèvre, A. M., & Eber, F. (1993). Interspecific hybrids between a transgenic rape-seed (*Brassica napus*) and related species: Cytogenetical characterization and detection of the transgene. *Genome, 36*, 1099–1106.

Kerlan, M. C., Chèvre, A. M., Eber, F., Baranger, A., & Renard, M. (1992). Risk assessment of outcrossing of transgenic rapessed to related species. *Euphytica, 62*, 145–153.

Kimber, D. S., & McGregor, D. I. (1995). *Brassica oilseeds: Production and utilization*. Cab international.

Knispel, A. L., & McLachlan, S. M. (2010). Landscape-scale distribution and persistence of genetically modified oilseed rape (*Brassica napus*) in Manitoba, Canada. *Environmental Science and Pollution Research, 17*, 13–25.

Knispel, A. L., McLachlan, S. M., Van Acker, R. C., & Friesen, L. F. (2008). Gene flow and multiple herbicide resistance in escaped canola populations. *Weed Science, 56*, 72–80.

Körber-Grohne, U. (1995). *Nutzpflanzen in Deutschland von der Vorgeschichte bis heute*. Stuttgart: Theiss Verlag.

Kumar, A., Premi, O. P., & Thomas, L. (2007). *Rapeseed-Mustard cultivation in India—An overview*. National Research Centre on Rapeseed, Mustard, Bharatpur 321 303 (Rajasthan).

Lamb, R. J. (1989). Entomology of oilseed Brassica crops. *Annual Review of Entomology, 34*, 211–229.

Landbo, L., Andersen, B., & Jørgensen, R. B. (1996). Natural hybridisation between oilseed rape and a wild relative: Hybrids among seeds from weedy *B. campestris*. *Hereditas, 125*, 89–91.

Langhof, M., & Rühl, G. (2017). Coexistence in oilseed rape: Effect of donor variety type and discarding field edges. *Journal of Agricultural Science, 9*, 33.

Lavigne, C., Klein, E. K., Vallée, P., Pierre, J., Godelle, B., & Renard, M. (1998). A pollen-dispersal experiment with transgenic oilseed rape. Estimation of the average pollen dispersal of an individual plant within a field. *Theoretical and Applied Genetics, 96*, 886–896.

Leterme, P. (1985). Modélisation de la croissance et de la production des siliques chez le colza d'hiver application à l'interprétation de résultats de rendements. Thèse de Docteur-Ingénieur de l'Institut National Agronomique Paris-Grignon.

Londo, J. P., Bautista, N. S., Sagers, C. L., Lee, E. H., & Watrud, L. S. (2010). Glyphosate drift promotes changes in fitness and transgene gene flow in canola (brassica napus) and hybrids. *Annals Of Botany, 106*, 957–965.

Lutman, P. J. W. (1993). The occurrence and persistence of volunteer oilseed rape (*Brassica napus*). *Aspects of Applied Biology, 35*, 29–36.

Männer, K. (2000). Haltbarkeit von pflanzlichen Verbreitungseinheiten nach Magen-Darm-Passage. UBA-Texte 19/00, Berlin.

Maxwell, P. S., Pitt, K. A., Olds, A. D., Rissik, D., & Connolly, R. M. (2015). Identifying habitats at risk: Simple models can reveal complex ecosystem dynamics. *Ecological Applications, 25*, 573–587. https://doi.org/10.1890/14-0395.1.

Mayer, M., Wurtz, A., Jülich, R., Roller, G., & Tappeser, B. (1995). *Anforderungen an die Überwachung von Freisetzungen gentechnisch veränderter Pflanzen und Mikroorganismen als Landesaufgabe im Rahmen des Vollzugs des Gentechnikgesetzes*. Naturschutz und Raumordnung, Sachsen-Anhalt: Ministerium für Umwelt.

McCartney, H. A., & Lacey, M. E. (1991). Wind dispersal of pollen from crops of oilseed rape (*Brassica napus* L.). *Journal of Aerosol Science, 22*, 467–477.

Mccauley, R., Davies, M. & Wyntje, A. (2012). The step-wise approach to adoption of genetically modified (gm) canola in Western Australia.

Menzel, G. (2006). Verbreitungsdynamik und Auskreuzungspotential von *Brassica napus* L.(Raps) im Großraum Bremen. GCA-Verlag, Waabs. ISBN 3-89863-213-X.

Menzel, G., Born, A. (2004). Abschnitt B2: Mittlere Darstellungsebene: Empirische Analyse der Verbreitung von Raps und Kreuzungspartnern in Bremen und im Bremer Umland. In Der Forschungsverbund GenEERA (Generische Erfassung Und Extrapolation Der Raps-Ausbreitung). Abschlussbericht and Das Bundesministerium Für Bildung Und Forschung. Universität Bremen.

Mesquida, J., Renard, M., & Pierre, J.-S. (1988). Rapeseed (*Brassica napus* L.) Productivity: The effect of honeybees (*Apis mellifera* L.) and different pollination conditions in cage and field tests. *Apidologie, 19*, 51–72.

Messean, A., Angevin, F., Go´mez-Barbero, M., Menrad, K., & Rodríguez-Cerezo, E. (2006). *New case studies on the coexistence of GM and non-GM crops in European agriculture*. JRC-IPTS-ESTO Technical report series EUR 22102 EN.

Messeguer, J., Peñas, G., Ballester, J., Bas, M., Serra, J., Salvia, J., et al. (2006). Pollen-mediated gene flow in maize in real situations of coexistence. *Plant Biotechnology Journal, 4*, 633–645.

Metz, P. L. J., Jacobsen, E., Nap, J. P., Pereira, A., & Stiekema, W. J. (1997). The impact on biosafety of the phosphinothricin-tolerance transgene in inter-specific *B. rapa* × *B. napus* hybrids and their successive backcrosses. *Theoretical and Applied Genetics, 95*, 442–450.

Middelhoff, U., Reuter, H., & Breckling, B. (2011). GeneTraMP, a spatio-temporal model of the dispersal and persistence of transgenes in feral, volunteer and crop plants of oilseed rape and related species. *Ecological Indicators, 11*, 974–988.

Mikkelsen, T. R., Andersen, B., & Jørgensen, R. B. (1996). The risk of crop transgene spread. *Nature, 380*, 31.

Mizuguti, A., Yoshimura, Y., Shibaike, H., & Matsuo, K. (2011). Persistence of feral populations of *Brassica napus* originated from spilled seeds around the Kashima seaport in Japan. *Japan Agricultural Research Quarterly: JARQ, 45*, 181–185.

Moser, D., Eckerstorfer, M., Pascher, K., Essl, F., & Zulka, K. P. (2013). Potential of genetically modified oilseed rape for biofuels in Austria: Land use patterns and coexistence constraints could decrease domestic feedstock production. *Biomass and Bioenergy, 50*, 35–44. https://doi.org/10.1016/j.biombioe.2012.10.004.

National Academies of Sciences. (2016). Gene drives on the horizon: Advancing science, navigating uncertainty, and aligning research with public values. The National Academies Press, Washington, DC. https://doi.org/10.17226/23405

Neemann, G., & Scherwaß, R. (1999). Materialen für ein Konzept zum Monitoring von Umweltwirkungen gentechnisch veränderter Pflanzen. UBA-Texte 52/99, Berlin.

Nishizawa, T., Nakajima, N., Aono, M., Tamaoki, M., Kubo, A., & Saji, H. (2009). Monitoring the occurrence of genetically modified oilseed rape growing along a Japanese roadside: 3-year observations. *Environmental Biosafety Research, 8*, 33–44.

Noble, C., Min, J., Olejarz, J., Buchthal, J., Chavez, A., Smidler, A. L., DeBenedictis, E. A., Church, G. M., Nowak, M. A., & Esvelt, K. M. (2016). *Daisy-chain gene drives for the alteration of local populations*.

Norris, C., & Sweet, J. (2002). Monitoring large scale releases of genetically modified crops (EPG 1/5/84): Incorporating report on project EPG 1/5/30–monitoring releases of genetically modified crop plants. *DEFRA Report, EPG, 1*, 84.

Norris, C. E., Simpson, E. C., Sweet, J. B., Thomas, J. E. (1999). Monitoring weediness and persistence of genetically modified oilseed rape (*Brassica napus*) in the UK. *Monograph-British Crop Protection Council*, 255–260.

OECD. (1997). Consensus document on the biology of *Brassica napus* L. (oilseed rape), OECD, Series on harmonization of regulatory oversight in biotechnology.

Organisation for Economic Co-operation and Development OECD. (2012). *Consensus document on the biology of brassica crops* (*Brassica* spp.).

Osborne, J. L., Clark, S. J., Morris, R. J., Williams, I. H., Riley, J. R., Smith, A. D., et al. (1999). A landscape-scale study of bumble bee foraging range and constancy, using harmonic radar. *Journal of Applied Ecology, 36*, 519–533.

Oye, K. A., Esvelt, K., Appleton, E., Catteruccia, F., Church, G., Kuiken, T., Lightfoot, S. B., McNamara, J., Smidler, A., & Collins, J. P. (2014). Regulating gene drives. *Science* (New York, N.Y.) *345*, 626–628. https://doi.org/10.1126/science.1254287

Pascher, K., & Dolezel, M. (2005). Koexistenz von gentechnisch veränderten, konventionellen und biologisch angebauten Kulturpflanzen in der österreichischen Landwirtschaft-Handlungsempfehlungen aus ökologischer Sicht. Forschungsauftrag des Bundesministeriums für Gesundheit und Frauen, Sektion IV., Bundesministerium für Gesundheit u. Frauen, Sekt. IV.

Pascher, K., & Gollmann, G. (1999). Ecological risk assessment of transgenic plant releases: An Austrian perspective. *Biodiversity & Conservation, 8*, 1139–1158.

Pascher, K., Hainz-Renetzeder, C., Gollmann, G., & Schneeweiss, G. M. (2017). Spillage of viable seeds of oilseed rape along transportation routes: Ecological risk assessment and perspectives on management efforts. *Frontiers in Ecology and Evolution, 5*, 104.

Pascher, K., Macalka, S., Rau, D., Gollmann, G., Reiner, H., Glössl, J., & Grabherr, G. (2010). Molecular differentiation of commercial varieties and feral populations of oilseed rape (*Brassica napus* L.). *BMC Evolutionary Biology, 10*, 63.

Pascher, K., Macalka-Kampfer, S., & Reiner, H. (2000). Vegetationsökologische und genetische Grundlagen für die Risikobeurteilung von Freisetzungen von transgenem Raps und Vorschläge für ein Monitoring, Bundesministerium f. soziale Sicherheit und Generationen, Forschungsberichte 7/2000.

Pascher, K., Narendja, F., Rau, D. (2006). *Feral oilseed rape: Investigations on its potential for hybridisation*; Final Report in Commission of the Federal Ministry of Health and Women (BMGH), Section IV; GZ: 70420/0116-IV/B/12/2004. Bundesministerium für Gesundheit u. Frauen.

Pekrun, C. (1994). Untersuchungen zur sekundären Dormanz bei Raps (*Brassica napus* L.). Cuvillier.

Pekrun, C., Lutman, P. J. W., & Baeumer, K. (1997). Germination behaviour of dormant oilseed rape seeds in relation to temperature. *Weed Research, 37*, 419–431.

Pekrun, C., Ripfel, H., Albertin, A., Lutman, P. J. W., & Claupein, W. (1998). Einfluß der Bodenbearbeitung auf die Ausbildung einer Samenbank bei Raps – Ergebnisse von sechs Standorten in England und einem in Österreich im Jahre 1997. *Mitteilungen der Gesellschaft für Pflanzenbauwissenschaften, 11*, 51–52.

Pessel, D., Lecomte, J., Emeriau, V., Krouti, M., Messean, A., & Gouyon, P. H. (2001). Persistence of oilseed rape (*Brassica napus* L.) outside of cultivated fields. *Theoretical and Applied Genetics, 102*, 841–846. https://doi.org/10.1007/s001220100583.

Pivard, S., Adamczyk, K., Lecomte, J., Lavigne, C., Bouvier, A., Deville, A., et al. (2008). Where do the feral oilseed rape populations come from? A large-scale study of their possible origin in a farmland area. *Journal of Applied Ecology, 45*, 476–485.

Prentis, P. J., Wilson, J. R., Dormontt, E. E., Richardson, D. M., & Lowe, A. J. (2008). Adaptive evolution in invasive species. *Trends in Plant Science, 13*, 288–294.

Rakow, G., & Woods, D. L. (1987). Outcrossing in rape and mustard under Saskatchewan prairie conditions. *Canadian Journal of Plant Science, 67*, 147–151.

Ramsay, G., Thompson, C., Squire, & G. R. (2003). Quantifying landscape-scale gene flow in oilseed rape. *Final Report of Defra Project* 1–50.

Ramsey, G., Thompson, C. E., Neilson, S., & Mackay G. R. (1999). Honeybees as vectors of GM oilseed rape pollen. *BCPC Symposium Proceedings* 72.

Raybould, A. F., & Gray, A. J. (1993). Genetically modified crops and hybridization with wild relatives: A UK perspective. *Journal of Applied Ecology* 199–219.

Reeves, R. G., & Reed, F. A. (2015). Stable transformation of a population and a method of biocontainment using haploinsufficiency and underdominance principles.

Reuter, H., Menzel, G., & Breckling, B. (2008). Hazard mitigation or mitigation hazard? *Environmental Science and Pollution Research, 15*, 529–535.

Rieger, M. A., Potter, T. D., Preston, C., & Powles, S. B. (2001). Hybridisation between *Brassica napus* L. and *Raphanus raphanistrum* L. under agronomic field conditions. *Theoretical and Applied Genetics, 103*, 555–560.

Ringdahl, E. A., McVetty, P. B. E., & Sernyk, J. L. (1987). Intergeneric hybridization of Diplotaxis spp. with *Brassica napus*: A source of new CMS systems? *Canadian Journal of Plant Science, 67*, 239–243.

Röbbelen, G. (1986). Anbau von 00-Raps in der Bundesrepublik Deutschland 1984/85. *Raps, 4*, 4–10.

Roller, A., Beismann, H., & Albrecht, H. (2001). Monitoring of the persistence of GM oil seed rape seeds in field trials. *Verhandlungen der GfÖ, 31*, 250.

Saji, H., Nakajima, N., Aono, M., Tamaoki, M., Kubo, A., Wakiyama, S., et al. (2005). Monitoring the escape of transgenic oilseed rape around Japanese ports and roadsides. *Environmental Biosafety Research, 4*, 217–222.

Salisbury, P. (1989). Potential utilization of wild crucifer germplasm in oilseed Brassica breeding. In *Proceeding ARAB 7th Workshop* (pp. 51–33), Toowoombu, Queensland, Australia.

Salisbury, P. A. (1991). *Genetic variability in Australian wild crucifers and its potential utilisation in oilseed Brassica species* (Ph.D.-thesis). La trobe University.

Sarwar, M. F., Sarwar, M. H., Sarwar, M., Qadri, N. A., & Moghal, S. (2013). The role of oilseeds nutrition in human health: A critical review. *Journal of Cereals and Oilseeds, 4*, 97–100.

Saure, C., Kühne, S., & Hommel, B. (1999). *Untersuchungen zum Pollentransfer von transgenem Raps auf verwandte Kreuzblütler durch Wind und Insekten*. Presented at the Proceedings zum BMBF-Statusseminar (p. 30).

Saure, C., Kühne, S., & Hommel, B. (1999). Auswirkung des Anbaus gentechnisch veränderter Rapspflanzen auf blütenbesuchende Bienen (Apidae) und Schwebfliegen (Syrphidae). Jahresbericht der BBA 1999, Berlin und Braunschweig.

Sauremann, W. (1987). Entwicklung des Glucosinolatgehaltes von 00-Raps in den nächsten 10 Jahren in Abhängigkeit vom Fremddurchwuchs. *Raps, 1*, 12–13.

Schafer, M. G., Ross, A. A., Londo, J. P., Burdick, C. A., Lee, E. H., Travers, S. E., et al. (2011). The establishment of genetically engineered canola populations in the US. *PLoS ONE, 6*, e25736.

Scheffler, J. A., & Dale, P. J. (1994). Opportunities for gene transfer from transgenic oilseed rape (*Brassica napus*) to related species. *Transgenic Research, 3*, 263–278.

Scheffler, J. A., Parkinson, R., & Dale, P. J. (1993). Frequency and distance of pollen dispersal from transgenic oilseed rape (*Brassica napus*). *Transgenic Research, 2*, 356–364.

Schlink, S. (1998a). 10 Jahre Überdauerung von Rapssamen (*Brassica napus* L.) im Boden. *Mitteilungen der Gesellschaft für Pflanzenbauwissenschaften, 11*, 221–222.

Schlink, S. (1998b). 10 years survival of rape seed (*Brassica napus* L.) in soil. *ZEITSCHRIFT FUR PFLANZENKRANKHEITEN UND PFLANZENSCHUTZ-SONDERHEFT-, 16*, 169–172.

Schlink, S. (1994). *Ökologie der Keimung und Dormanz von Körnerraps (Brassica napus L.) und ihre Bedeutung für eine Überdauerung der Samen im Boden*. Dissertationes Botanicae.

Schmidt, G., & Schröder, W. (2011). Regionalisation of climate variability used for modelling the dispersal of genetically modified oil seed rape in Northern Germany. *Ecological Indicators, 11*, 951–963.

Schoenenberger, N., & D'Andrea, L. (2012). Surveying the occurrence of subspontaneous glyphosate-tolerant genetically engineered *Brassica napus* L. (Brassicaceae) along Swiss railways. *Environmental Sciences Europe*, 24, 23.

Schulze, J., Brodmann, P., Oehen, B., & Bagutti, C. (2015). Low level impurities in imported wheat are a likely source of feral transgenic oilseed rape (*Brassica napus* L.) in Switzerland. *Environmental Science and Pollution Research, 22*, 16936–16942.

Schulze, J., Frauenknecht, T., Brodmann, P., & Bagutti, C. (2014). Unexpected diversity of feral genetically modified oilseed rape (*Brassica napus* L.) despite a cultivation and import ban in Switzerland. *PLoS One 9*, e114477.

Scott, S., & Wilkinson, M. (1998). Transgenic risk is low. *Nature, 393*, 320.

Sikka, Sm. (1940). Cytogenetics of Brassica hybrids and species. *Journal of Genetics, 40*, 441–474.

Simard, M.-J., Légère, A., Séguin-Swartz, G., Nair, H., & Warwick, S. (2005). Fitness of double versus single herbicide–Resistant canola. *Weed Science, 53*, 489–498.

Simard, M.-J., Légère, A., & Warwick, S. I. (2006). Transgenic *Brassica napus* fields and *Brassica rapa* weeds in Quebec: Sympatry and weed-crop in situ hybridization. *Botany, 84*, 1842–1851.

Simpson, E. C., Norris, C. E., Law, J. R., Thomas, J. E., & Sweet, J. B. (1999). Gene flow in genetically modified herbizide tolerant oilseed rape (*Brassica napus*) in the UK. *BCPC Symposium Proceedings, 72*.

Snow, A. A. (2002). Transgenic crops—Why gene flow matters. *Nature Biotechnology, 20*, 542.

Sobrino-Vesperinas, E. (1988). Obtainment of some new intergeneric hybrids between wild Brassiceae. *Candonella, 43*, 499–504.

Song, K. M., Osborn, T. C., & Williams, P. H. (1990). Brassica taxonomy based on nuclear restriction fragment length polymorphisms (RFLPs) 3. Genome relationships in Brassica and related genera and the origin of *B. oleracea* and *B. rapa* (syn. campestris). *Theoretical and Applied Genetics, 79*, 497–506.

Squire, G., Crawford, J., Ramsey, G., & Thompson, C. E. (1999). *BCPC Symposium Proceedings, 72*, 57–64.

Squire, G. R., Breckling, B., Pfeilstetter, A. D., Jorgensen, R. B., Lecomte, J., Pivard, S., et al. (2011). Status of feral oilseed rape in Europe: Its minor role as a GM impurity and its potential as a reservoir of transgene persistence. *Environmental Science and Pollution Research, 18*, 111–115.

Stace, C. (2010). *New flora of the British Isles*. Cambridge University Press.

Tamis, W. L. M., & De Jong, T. J. (2010). *Transport chains and seed spillage of potential GM crops with wild relatives in the Netherlands* (pp. 2010–2102). COGEM report: CGM.

Thomas, D. D., Donnelly, C. A., Wood, R. J., & Alphey, L. (2000). Insect population control using a dominant, repressible, lethal genetic system. *Science, 287*, 2474–2476.

Timmons, A. M., O'Brien, E. T., Charters, Y. M., Dubbels, S. J., & Wilkinson, M. J. (1995). Assessing the risks of wind pollination from fields of genetically modified *Brassica napus* ssp. oleifera. In *The methodology of plant genetic manipulation: Criteria for decision making* (pp. 417–423). Springer.

Treu, R., & Emberlin, J. (2000). Pollen dispersal in the crops Maize (Zea mays), Oil seed rape (*Brassica napus* ssp. oleifera), Potatoes (Solanum tuberosum), Sugar beet (Beta vulgaris ssp. vulgaris) and Wheat (Triticum aestivum). Evidence from publications. A report for the Soil Association from the National Pollen Research Unit, University College Worcester.

United States Department of Agriculture—USDA (2018). *Oilseeds: World markets and trade.*

von der Lippe, M., & Kowarik, I. (2007). Crop seed spillage along roads: A factor of uncertainty in the containment of GMO. *Ecography, 30*, 483–490.

Vullioud, P. (1992). Densité de semis en culture de colza d'automne. *Rev Suisse Agric, 24*, 345–350.

Waddington, K. D., Herbert, T. J., Visscher, P. K., & Richter, M. R. (1994). Comparisons of forager distributions from matched honey bee colonies in suburban environments. *Behavioral Ecology and Sociobiology, 35*, 423–429.

Warwick, S. I. (1997). *Use of biosystematic data, including molecular phylogenies, for biosafety evaluation*. Presented at the JIRCAS International Symposium Series (Japan).

Warwick, S. I., Legere, A., SIMARD, M.-J., James, T. (2008) Do escaped transgenes persist in nature? The case of an herbicide resistance transgene in a weedy *Brassica rapa* population. *Molecular Ecology, 17*, 1387–1395

Warwick, S. I., Simard, M.-J., Légère, A., Beckie, H. J., Braun, L., Zhu, B., et al. (2003). Hybridization between transgenic *Brassica napus* L. and its wild relatives: *Brassica rapa* L., *Raphanus raphanistrum* L., *Sinapis arvensis* L., and *Erucastrum gallicum* (Willd.) OE Schulz. *Theoretical and Applied Genetics, 107*, 528–539.

Wilkinson, M. J., Timmons, A. M., Charters, Y., Dubbels, S., Robertson, A., Wilson, N., Scott, S., O'Brien, E., Lawson, H. M. (1995). *Problems of risk assessment with genetically modified oilseed rape*. Presented at the Brighton crop protection conference: Weeds. Proceedings of an international conference, Brighton, UK, 20–23 November 1995. British Crop Protection Council, pp. 1035–1044.

Yoshimura, Y., Beckie, H. J., & Matsuo, K. (2006). Transgenic oilseed rape along transportation routes and port of Vancouver in western Canada. *Environmental Biosafety Research, 5*, 67–75.

Zander, P. (2003). *Agricultural land use and conservation options—A modelling approach.* Dissertation, Wageningen University. 240.

6

Gene Drives and Population Dynamics: Model Concepts

Johannes L. Frieß, Merle Preu and Broder Breckling

Introduction

Recently, the discussion on gene drives is gaining momentum (Fischer 2018; Groß 2016; Jacobs 2015). The deliberate release of genetically modified organisms (GMOs) equipped with genetic systems to overcome the patterns of Mendelian inheritance, able to either drive desired traits into a wild population or suppress the population altogether. This harbours wide implications concerning economy, ethics, genetics and ecology. To explore implications of the latter we constructed a series of model concepts around the putative target organism, the olive fruit fly *Bactrocera oleae*. We discuss the different modelling approaches in their current prototype stages and want to stress the importance of model simulations in order to identify ecological hazards prior to the release of potentially adverse agents such as artificial self-reproducing.

The GeneTip project works on the conception and design modelling of population dynamics influenced by gene drives. In this pursuit, multiple different approaches and concepts have been developed to on one hand, be able to cover a broad perspective on the topic but on the other hand to also focus on different key aspects. In the following we present seven concepts based on different modelling approaches. In these model concepts our model organism is the olive fruit fly (*Bactrocera oleae*), which is a major pest species in agricultural olive production.

Female olive flies oviposit their eggs into the ripening olive fruits right beneath the outer skin (Nardi et al. 2005), thereby they lay one egg per fruit and are able lay

J. L. Frieß (✉)
Institute of Safety/Security and Risk Sciences, University of Natural Resources and Life Sciences, Vienna (BOKU), Vienna, Austria
e-mail: johannesfriess@gmx.net

M. Preu · B. Breckling
Chair of Landscape Ecology, University of Vechta, Vechta, Germany

250–350 eggs within a lifetime (Genç and Nation 2008). From an egg, a monodietary larva hatches which eats its way through the fruit pulp to the center of the olive (Daane and Johnson 2010; Sharaf 1980). Olive flies may have three to five overlapping generations per year (Bocaccio and Petacchi 2009; Comins and Fletcher 1988; Kokkari et al. 2017; Pontikakos et al. 2010; Voulgaris et al. 2013) and thereby cause annual net losses estimated to be as high 5% of global olive production translating into 800 million US-Dollar per year (Montiel Bueno and Jones 2002). Due to their role as an agricultural pest species, the olive fly is considered a target species for diverse pest control methods.

Until now they have been the target for the sterile insect technique (SIT) (Knipling 1955). Wherein, laboratory-reared sterile males are mass-released into the wild. The sterile males outnumber the fertile wild males and thus reduce the number of offspring in the subsequent generation. However, prolonged SIT applications proved costly and showed only limited success, since the decimated population rebounds quickly to their former size once the sterile male releases seize. Therefore, the application of a gene drive might be feasible (For a more comprehensive information on the olive fly review Chap. 4, case study 1).

A gene drive is a phenomenon that increases the prevalence of certain gene, or set of genes within a population. Artificial gene drive systems may be released with the intention to spread desired genes in a population, these genes may either confer a certain trait such as reduced fertility or a bias in the sex ratio of a population, but may also simply kill potential offspring. Such gene drives, may however harbour unforeseen ecological consequences, due to the multifactorial set-up of ecosystems. For instance, a potent gene drive aimed at suppression might cause an eradication of the target species which may affect interconnected species and the ecosystem's food web. Therefore, it is important to predict potential effects of gene drives on a population and their habitat by various modelling approaches.

First, the possibilities to model population dynamics based on a stock-flow model will be explored. This model will consider the temperature dependent mortality of the different life stages of the flies. Another stock flow model was developed to simulate the population dynamics of a single-locus Underdominance gene drive as presented by Reeves et al. (2014). Then a differential-equation based concept which further considers a gene drive released into a wild population of olive flies will be introduced. The following stochastic, recurrence equation-based model will further consider the variability enhancing effects of ecological disturbances such as winter population bottlenecks on an olive fly population that is already exposed to a gene drive. Finally, this chapter will conclude with an outlook onto an individual-based model which also considers the spatio-temporal dynamics of an olive fly population exposed to gene drive-carrying flies.

Stock-Flow Model of an Olive Fly Population

The stock flow model was put together with the isee systems STELLA® Professional Version 1.1.2 software. In its current state, it considers the temperature-dependent mortality data from the literature (Genç and Nation 2008; Sánchez-Ramos et al. 2013; Tsiropoulos 1972; Tsitsipis 1980) compared to the average annual temperatures encountered on the Greek island of Crete between 1961 and 1990 (German Weather Service).

This model (depicted in Fig. 6.1) was conceived as a means to visualize the growth and development of an olive fly population on an olive orchard. As datasets for this approach the collated data from the literature on *B. oleae* were used. In its current form, each life stage is integrated as a stock. The life cycle begins in the "Egg"-stock. The egg stock is limited by an estimated number of 100 million olives on the orchard. Influenced by the operator "Egg mortality", each egg may either die or follow the "Hatching"-flow to the next stage, depicted in the "Larva"-stock. From there, it goes on over the "Pupa"-stock to the "Male Adult" and "Female Adult"-stocks. Until the adult stage, the sex of the organisms is unimportant. During each life stage, the individual may either die or proceed to the next life stage. Each life stage is represented by a different stock. The mortality rates of the different life stages

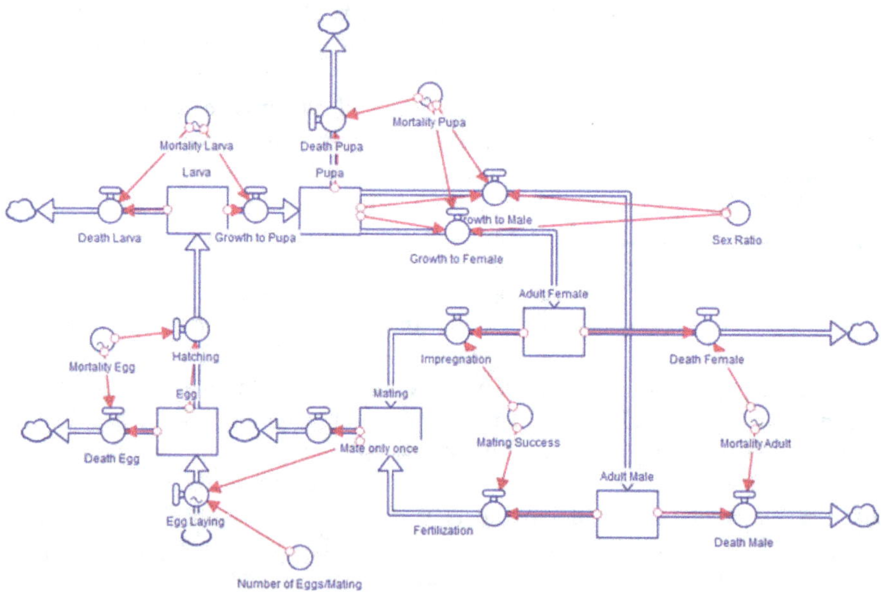

Fig. 6.1 Stock-flow model of an olive fly population. The flies undergo the different stages of egg, larva, and pupa, up to male and female adults. Each life stage is represented by a different stock. Male and female adults unite in the "Mating"-stock which stimulates the inflow into the "Egg"-stock. At each stage, the individual organism may either die due to temperature or transition to the next stage

in relation to the temperature as encountered in the cycle of the years on the Greek island of Crete were integrated into the model as the sole time dependent factor. The decision whether to follow the respective "Death" or "Growth"-flow is influenced by the "Mortality"-Operators. Note, that in its current form, the durations of the different life stages all equal one month. The adult flies, again may either die or unite in the "Mating"-stock. Since the "Mating"-stock's only outflow "Mate only once" leads to a sink, the model assumes a single mating per adult fly. The number of individuals in the "Mating"-stock stimulate the "Egg-Laying"-inflow into the "Egg"-stock. The "Number of Eggs/Mating"-operator adds a stochastic factor, as each mating produces between 150 and 400 eggs. Figure 6.2 shows a simulation run for 60 months. The initial population was set to contain only 1000 male- and 1000 female adult flies. From which it develops quickly and before the end of the second year the number of eggs for the first time reaches the maximum threshold of 100 million, which was set as the number of olives available for oviposition. Clear annual cycles are visible in the various life stage populations which are due to the variable temperature

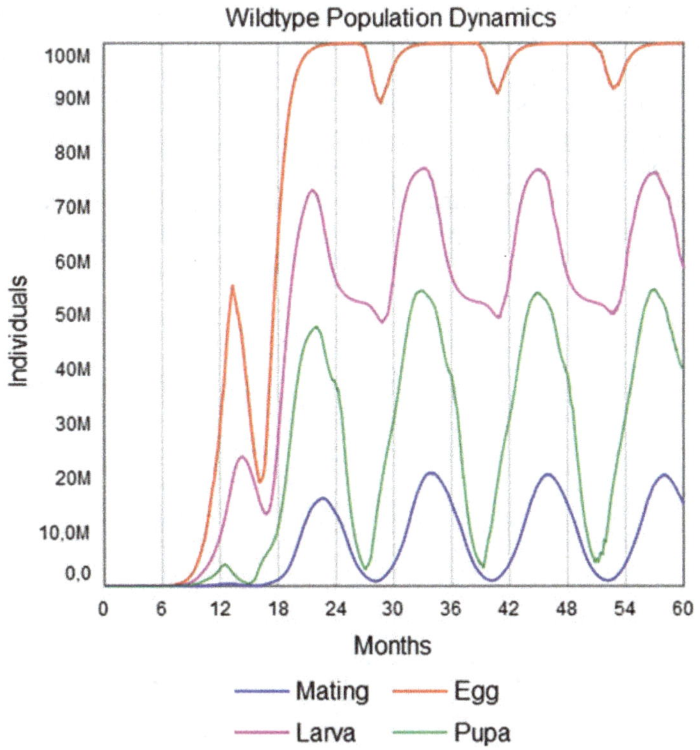

Fig. 6.2 Wild type population dynamics over a simulated time of 60 months. Note, that in this simplified model, the duration of each life stage equals one month. The maximum number of eggs is limited by the number of olives, which were set to 100 million. The population started with 1000 male and 1000 female adult flies. The life stage populations seem to reach stable annual cycles

dependent mortality rates. The population seems to undergo stable cycles. Further planned implementations into the model entail variable life stage durations, multiple matings per generation and a seasonal change in the number of olives available for oviposition.

The next step for this model would be to include applications of different gene drive techniques in order to examine the effects on the population dynamics. The gene drive techniques planned to implement are Medea (Akbari et al. 2014), two-locus Underdominance (Akbari et al. 2013), Y-linked X-Shredder (Galizi et al. 2014), CRISPR/Cas-based gene drives (Burt et al. 2018; Esvelt et al. 2014; Hammond et al. 2016) and Killer-Rescue (Gould et al. 2008).

Stock-Flow Model of a Single-Locus Underdominance Gene Drive

Underdominance, aka heterozygosity inferiority is a phenomenon in which heterozygous carriers of an underdominant gene suffer a higher fitness penalty than homozygous carriers. Modelling the population dynamics of a single-locus Underdominance gene drive requires the implementation of three genotypic subpopulations, wild types (W/W), homozygous (W/U) and heterozygous gene drive-carriers (U/U). These subpopulations are depicted as stocks as illustrated in Fig. 6.3.

This simple model only assumes static growth by a fixed growth variable, determining the number of offspring per mating. However, the number of gene drive

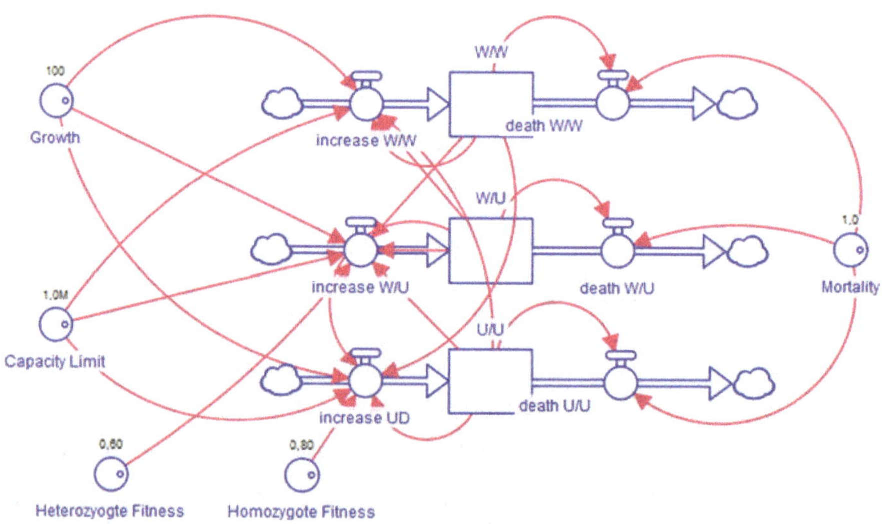

Fig. 6.3 Model structure of the stock-flow model for single-locus Underdominance gene drive population dynamics

carrying offspring is further affected by the respective fitness variables, included in the form of operators. Furthermore, growth of all subpopulations is limited by a universal capacity limit. Besides growing, the subpopulations can of course also decline due to mortality which was assumed equal for all subpopulations in this model. A sex ratio of 1:1 and random mating are assumed and therefore the number and genotypes of offspring are directly dependent on the size of the respective subpopulations. Adhering to the inheritance scheme of single-locus Underdominance, the Mendelian genotype ratios are derived. From which the following equations for the growth of each subpopulation are deductible.

$$P_{W/W} = +\frac{\left(N_{W/W} * N_{W/W}\right) + \left(\frac{1}{2} * N_{W/U} * N_{W/W}\right) + \left(\frac{1}{4} * N_{W/U} * N_{W/U}\right)}{N_{W/W} + N_{W/U} + N_{U/U}} \tag{6.1}$$

$$P_{W/U} = +\frac{\left(\frac{1}{2} * N_{W/U} * N_{W/U}\right) + \left(\frac{1}{2} * N_{W/U} * N_{U/U}\right) + \left(\frac{1}{2} * N_{W/U} * N_{W/W}\right) + \left(\frac{1}{2} * N_{W/W} * N_{U/U}\right)}{N_{W/W} + N_{W/U} + N_{U/U}} \tag{6.2}$$

$$P_{U/U} = +\frac{\left(N_{U/U} * N_{U/U}\right) + \left(\frac{1}{2} * N_{\frac{W}{U}} * N_{U/U}\right) + \left(\frac{1}{4} * N_{\frac{W}{U}} * N_{W/U}\right)}{N_{W/W} + N_{W/U} + N_{U/U}} \tag{6.3}$$

where $P_{X/X}$ represent the probabilities of offspring with the certain genotype and $N_{X/X}$ the number of individuals with that genotype in the population. This model is suited to examine the invasiveness of a single-locus Underdominance gene drive, without the affliction of ecological factors.

Stock-Flow Model of a Medea Gene Drive

Similar but more complex than Underdominance in its inheritance, the Medea gene drive technique can be simulated in a stock-flow model. The model structure is depicted in Fig. 6.4. Just as in the Underdominance model each genotypic subpopulation is represented as a stock. Furthermore, the different sexes become important as Medea's lethality is induced maternally. Hence, the model revolves around six stocks representing both sexes of three different genotypes. Each genotypic subpopulation's growth is regulated by a respective "Birth Rate"-converter, while the mortality of all subpopulations is regulated by a single "Mortality"-converter set to 1. The birth rate is set to 100 offspring per mating. As before with Underdominance, mating is assumed to be random and dependent on the size of the respective subpopulations. Obviously, male and female mating partners are required for reproduction, thus the number of mating events is equal to the number of the rarer mating partner (i.e. 50 WT males and 25 WT females will amount to 25 matings).

The populations' growth is also limited by the growth capacity already defined in the previous model in Eq. 6.4. In this model as well the capacity limit K was set to

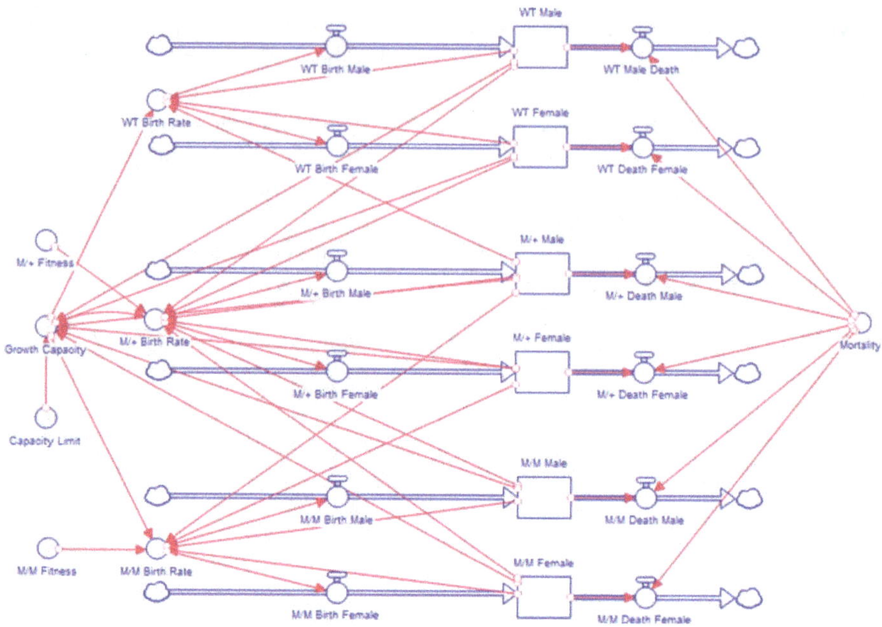

Fig. 6.4 Model structure of the stock-flow model representing Medea-gene drive population dynamics

10 million. The carrier subpopulations are furthermore affected in their growth by their respective fitness-converters. The homozygotes' fitness was assumed as 0.35 and the heterozygotes' fitness as 0.72, according to Buchman et al. (2018). The most important factor in the propagation of the gene drive is of course based on genetics. Only the offspring of a heterozygous Medea-carrying mother that do not possess a copy of the gene drive will die. Out of 36 different gamete combinations three are lethal, six are wild type, eighteen are heterozygous and nine are homozygous carriers.

From the inheritance scheme, the growth for each subpopulation can be deducted as shown in the following equations (Eqs. 6.4–6.6) and included in the model into the respective "Birth Rate"-converters. Similar to the model concept of Underdominance, this model is suited to simulate the invasiveness of Medea gene drives.

$$P_{+/+} = P(N^{\text{♀}}_{+/+} * N^{\text{♂}}_{+/+}) + \frac{1}{2} P(N^{\text{♂}}_{M/+} * N^{\text{♀}}_{+/+}) \tag{6.4}$$

$$\begin{aligned} P_{M/+} = {} & \frac{1}{2} P(N^{\text{♀}}_{M/+} * N^{\text{♂}}_{+/+}) + \frac{1}{2} P\left(N^{\text{♀}}_{+/+} * N^{\text{♂}}_{M/+}\right) + \frac{1}{2} P(N^{\text{♀}}_{M/+} * N^{\text{♂}}_{M/+}) + \\ & P(N^{\text{♀}}_{M/M} * N^{\text{♂}}_{+/+}) + P(N^{\text{♀}}_{+/+} * N^{\text{♂}}_{M/M}) + \frac{1}{2} P(N^{\text{♀}}_{M/+} * N^{\text{♂}}_{M/M}) + \frac{1}{2} P(N^{\text{♀}}_{M/M} * N^{\text{♂}}_{M/+}) \end{aligned} \tag{6.5}$$

$$P_{M/M} = P\left(N^{\female}_{M/M} * N^{\male}_{M/M}\right) + \frac{1}{2}P\left(N^{\female}_{M/+} * N^{\male}_{M/M}\right) + \frac{1}{2}P\left(N^{\female}_{M/M} * N^{\male}_{M/+}\right) + \frac{1}{4}P\left(N^{\female}_{M/+} * N^{\male}_{M/+}\right) \quad (6.6)$$

As before, $P_{X/X}$ denotes the probability of a certain offspring genotype, while $N_{X/X}$ is the number of genotypic individuals within the population, in this model divided by sex.

Deterministic Recurrence-Based Calculations on the Inheritance Schemes of Different Gene Drive Techniques

The model is based on the inheritance schemes of the various gene drive techniques. The probability of the occurrence of a particular genotype was multiplied by its respective fitness. A random mating was assumed based on the respective genotypes in the infinite population. The genotype fitness and the initial population share can be selected by the user. To calculate the invasiveness of gene drive techniques, the variation in fitness and population parameters for a given generation was automated and presented in a color-coded graph in which population percentages are assigned different colours depending on the thresholds selected. The purpose of this survey is to describe the ability of gene drives to achieve population replacements. For this purpose, a simple population genetic model was chosen that provides the percentage of genotype as a function of fitness and the relative release size of gene drive organisms (GDOs). Models with a similar approach have also been described by Gould et al. (2008), Ward et al. (2011) and Dhole et al. (2018).[1]

As a basis, our underlying recurrence-based, deterministic model utilizes inheritance schemes in a hypothetical population of infinite size. Therefore, it is impossible to achieve a population suppression or eradication in this model. All genotypic subpopulations are regarded as relative percentages of the whole population.

Two-locus Underdominance, Medea, a CRISPR/Cas-mediated GD including resistance allele formation, Killer-Rescue and a Y-linked X-Shredder were chosen for a quantitative comparison of their invasiveness. As positive and negative controls, the spread of two different transgenes lacking the GD-specific functionality of super-Mendelian inheritance were calculated: a female specific release of insects carrying a dominant lethal (fsRIDL, Fu et al. 2010) a fitness gain conferring transgene e.g. a pesticide resistance, respectively. For the calculations, the following assumptions were made that should be taken into account for a critical discussion of the presented results on the invasiveness:

- **fsRIDL (negative control)**: Female lethality is 100% regardless of zygosity. Cumulative fitness penalty for each allele was assumed.

[1] This model was already published under Creative Commons license.

Frieß JL, von Gleich A, Giese B, 2019. Gene drives as a new quality in GMO releases - a comparative technology characterization. PeerJ 7:e6793.

- **Transgene with fitnessloss/gain (negative/positive control)**: Cumulative fitness loss/gain for each allele was assumed.
- **X-Shredder**: A Y-linked X-Shredder system with a male biased sex ratio of 95% was assumed according to Galizi et al. (2014). Since the ratio of females cannot decrease below 5%, due to the assumptions of our model, the thresholds were adapted for this approach to 7 and 93% in the cross section computation.
- **Killer-Rescue**: Cumulative fitness penalties were assumed per allele regardless of killer or rescue.
- **Medea**: Fitness penalty was assumed being cumulative for hetero- and homozygous Medea-carriers.
- **Underdominance**: A two-locus autosomal Underdominance system similar to UD^{mel} (Akbari et al. 2013) was modelled. Female-carriers kill offspring that do not inherit at least one copy of each construct. It was assumed that heterozygosity for each of the Underdominance alleles confers a 15% fitness penalty. Therefore, the double hetero UD genotype's fitness is 30% lower than that of the double homo genotype. Whereas homozygosity in one construct but lack of the other results in half the fitness penalty of the double homozygotes.
- **CRISPR/Cas-mediated gene drive**: The homing rate was assumed to be 98% similar to data presented by DiCarlo et al. (2015). Resistance formation rates were assumed as the direct reciprocals of the homing rates, i.e. 2%. Fitness penalties were assumed to be half for heterozygous GDOs. Each resistance allele was assumed to confer a 10% fitness penalty. Homozygously resistant population percentages above 95% are depicted in green in the overlays.

Figure 6.5 shows positive and negative control approaches for the model as transparent overlays of cross sections for up to 60 generations post release, in 5-generational steps. Negative controls are represented by the complete fading of fsRIDL-carriers from a population within five generations and the spread of a transgene which confers a fitness loss. The spread of a transgene which confers a fitness gain represents a positive control. The blue areas represent combinations of fitness

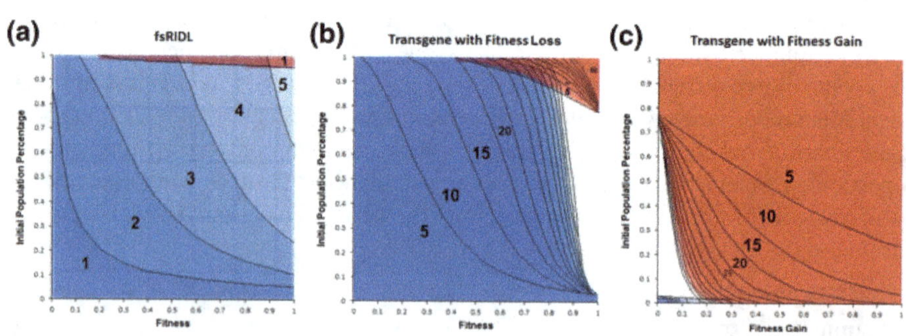

Fig. 6.5 Cross sections of the negative controls for the model represented on the fsRIDL technique and the fade of a transgene conferring a fitness loss. Red = Wild type population percentage below 5%; Blue = Wild type population percentage above 95%. Black numbers and lines represent the respective generation post-release. Lines were inserted by hand for clarity

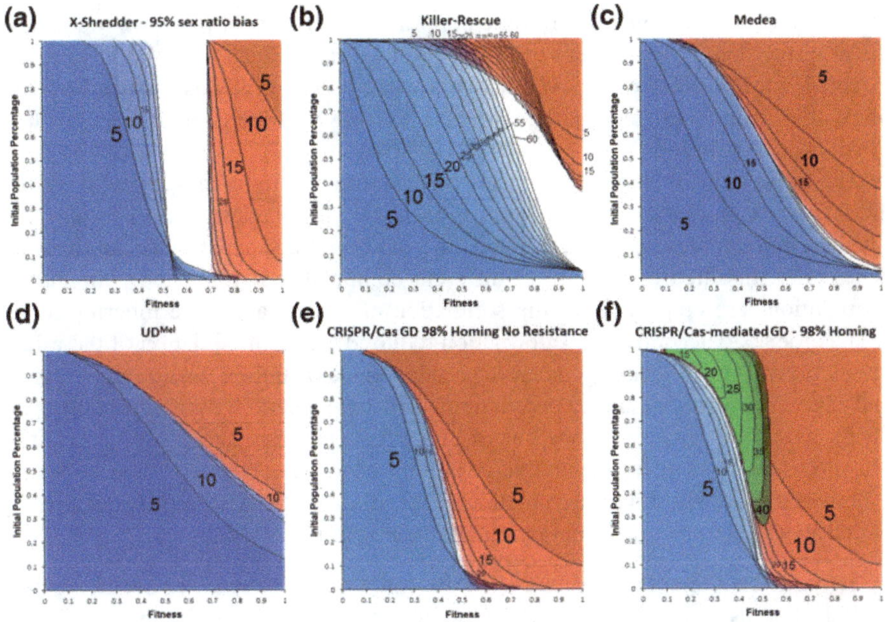

Fig. 6.6 Cross section overlays of fitness and initial release population percentage for different gene drive techniques. Red = Wild type population percentage below 5%; Blue = Wild type population percentage above 95%. Green = Population homozygous for resistance alleles above 95%. Black numbers and lines represent the respective generation post-release. Black lines were inserted by hand for clarity

and population percentage at a given generation post-release at which more than 95% of the population is of wild type genotype. Red areas represent fitness and population percentage combinations at given generations resulting in less than 5% wild type genotypes in the population.

Figure 6.6 shows the allele frequency of gene drives. The depicted generations post-release range from 5 to a maximum of 60, in 5-generational steps.

Stochastic Model Considering an Olive Fly Population with Gene Drive and Bottlenecks

For this model, a gene drive technique was assumed resembling a CRISPR/Cas-based gene drive carrying a female specific lethal cargo-gene, similar to the RIDL [Release of Insects carrying a Dominant Lethal gene (Thomas et al. 2000)]. Therefore, it was assumed that 99% of the offspring resulting from a mating of a wild type and an AG parent will also be altered gene individuals. This corresponds a conversion rate similar to findings reported by DiCarlo et al. in wild yeast (DiCarlo et al. 2015).

Furthermore, it is assumed that if these offspring are female, they are non-viable and die at larval stage. Mating between two wild type individuals will of course result in 100% wild type offspring. And lastly, a winter bottleneck was assumed to take place after every sixth generation, thereby killing 98% of the population regardless of genetic constitution.

The model was implemented in the statistical programming language "R", using the RLab and LaPlaces Demon Library packages. It was designed to display the size and demographic characteristics of the wild type *B. oleae* population and the genetically modified population at different computational time steps. Stochastic formulations are employed for the simulation of mating and gene inheritance, as well as for sex determination. Information required for the modulation of the initial population and the released gene drive population was acquired via literature review and expert surveys. The initial population consists of a certain number of wild type (WT) males and females, as well as of individuals carrying an altered gene (AG). The model assumes a certain percentage of females to mate, which are selected using a Bernoulli distribution for each mating event. In order to select a male for mating with a specific female, random mating and monogamy were assumed.

Two different simulation scenarios were run, one with regularly occurring bottlenecks and as a control a scenario without winter bottlenecks. In both scenarios, an initial wild population of 200,000 olive fruit flies (sex ratio 1:1) was assumed. A total of 400,000 gene drive-carrying males were released. Ten replicate simulation runs were performed; each comprises a simulation for 24 generations.

Results of these model scenarios are depicted in Fig. 6.7a and 23 Without the occurrence of a population bottleneck, the release of altered gene flies induces a constant reduction of population densities over the course of the first 11 simulated generations. Thereafter, the fly population recovers and increases from generation to generation. Figure 6.7b shows that a similar outcome is obtained for each simulation run. Figure 6.8 shows that the occurrence of a bottleneck after six generations further reduces the number of flies which is already on decline due to the release of gene drive-carrying individuals. While altered gene flies exhibit a high proportion of the entire population after the first bottleneck, wild type flies dominate the population after the second and third bottleneck. Very different outcomes are displayed by single model runs: in some cases, the wild type population is driven to extinction, in other model runs it recovers at various rates.

These results indicate that the occurrence of population bottlenecks reduces the predictability of the effect of gene drive releases to natural olive fruit fly populations. Therefore, the consideration of ecological properties at the population biological level for a comprehensive risk assessment prior to the release of genetically altered organisms to wild ecosystems is strongly recommended.

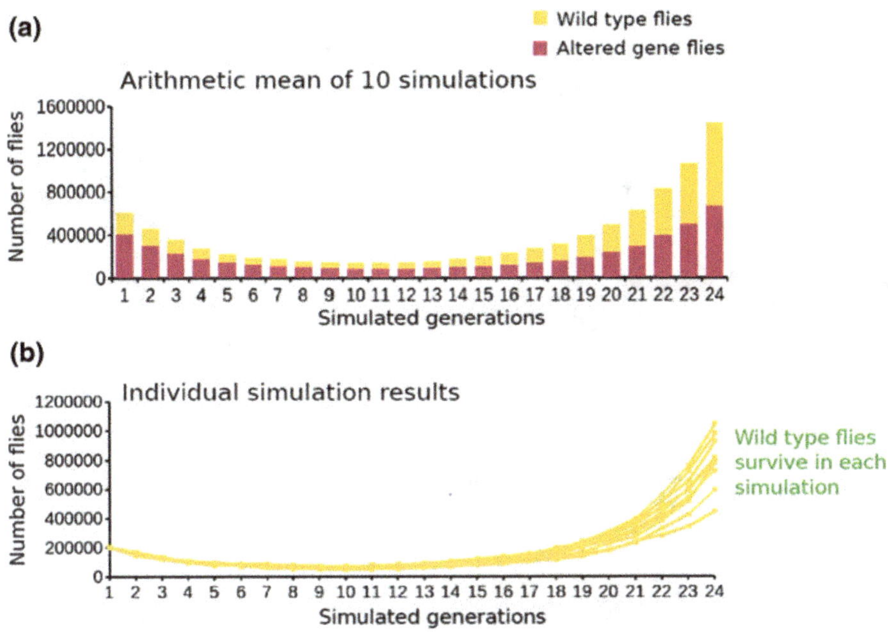

Fig. 6.7 Simulation result for an olive fruit fly population that does not experience population bottlenecks. **a** Arithmetic mean of ten model runs; **b** development of the wild type fly population in each single model run

Differential Equation-Based Modelling of an Olive Fly Population with a Gene Drive

Differential equations are frequently used in population ecology. They can be considered as a standard to study spatially homogeneous conditions, where the calculation of non-overlapping generations is not applicable. The results from both approaches can substantially differ, since the causality-derived determinism requires an absence of trajectories crossing, whereas in difference equations, the systems state is only defined for certain points in time.

For this deterministic approach, the following assumptions have been made: The population grows exponentially up to an environmentally set carrying capacity limit. The individuals of the population die following an exponential decline. Just as before, the sex ratio of the wild type population is assumed to be 1:1. The gene drive, similar to the one in the stochastic model, is assumed to be propagated to all offspring as long as one of the mating partners carries the gene drive and the gene drive causes female lethality.

For the two sub-populations the following equations were established:

$$\text{Wild type population}\ \frac{dW}{dt} = r * W * \frac{0.5 * W}{\frac{1}{2}W + Mg} * \frac{K - (W + Mg)}{K} - mf * FF * W \tag{6.7}$$

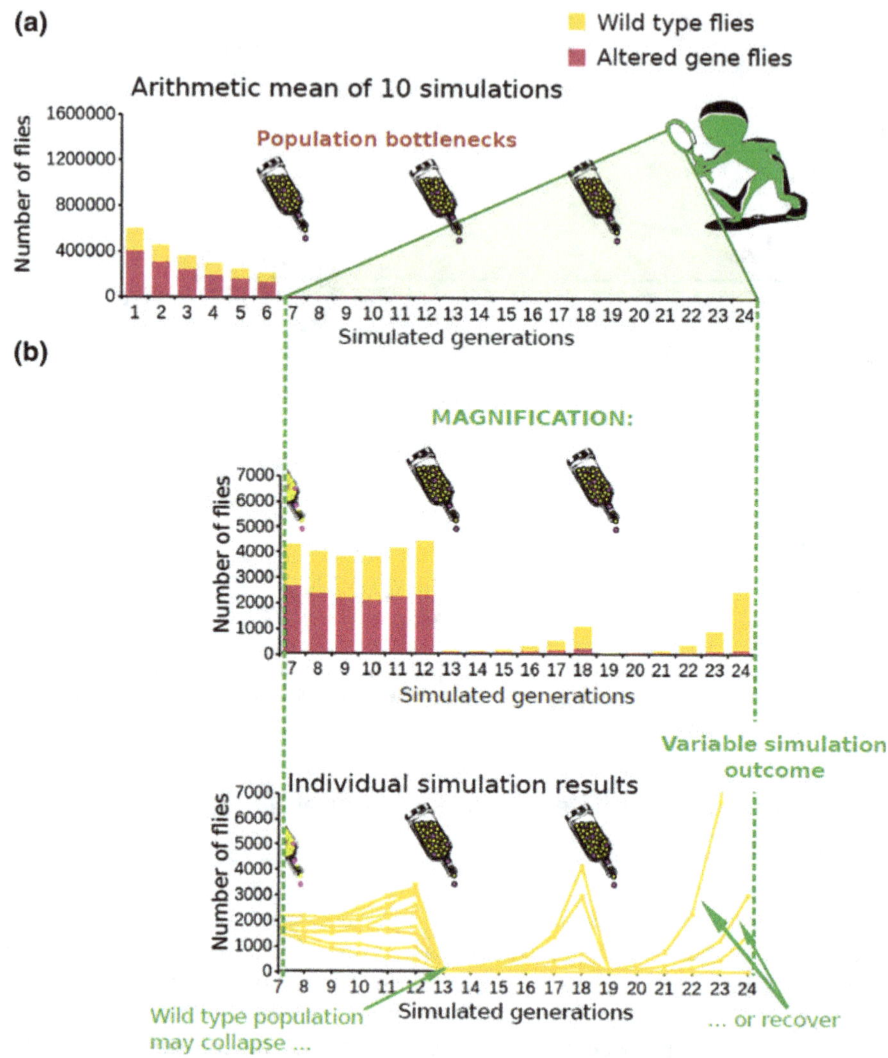

Fig. 6.8 Simulation result for an olive fruit fly population regularly experiencing population bottlenecks. **a** The arithmetic mean of ten model runs; **b** development of the wild type fly population for each single model run. The bottles indicate the regular occurrence of population bottlenecks after every sixth generation

$$\text{Male gene drive population } \frac{dMg}{dt} = r * W * \frac{Mg}{\frac{1}{2}W + Mg} * \frac{K - (W + Mg)}{K} - mf * Mg \quad (6.8)$$

where r denotes the exponential growth and is assumed to be 0.5; W is the population of the wild types, Mg is the population of the male gene drive-carriers; K is the carrying capacity of the ecosystem and is assumed to be 2000; mf is the mortality

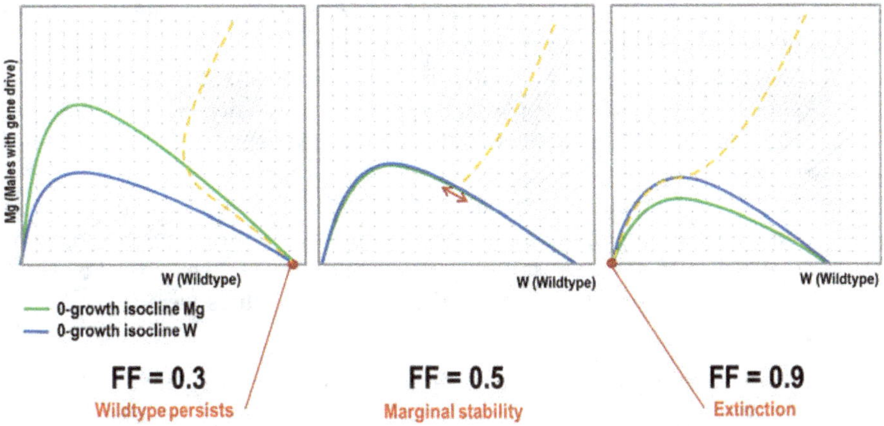

Fig. 6.9 Direction fields showing the wild type versus the male gene drive-carrying population. Dependent on the FF isoclines form, which regardless of the initial population settings lead to the population's extinction for a fitness factor (FF above 0.5 and to wild type persistence for FF < 0.5. A marginal stability can be observed for FF = 0.5. X- and Y axes scale from 0.1 to 2000

factor, assumed to be 0.1; and *FF* is the fitness factor which represents the fitness difference between the populations, it was varied during simulation runs between 0.1 and 0.9.

During the simulation runs in which not only the *FF* but also the initial population values *W* and *Mg* have been varied, an interesting phenomenon could be observed. The isoclines vary depending on the *FF* value. In Fig. 6.9, these isoclines are shown in direction fields. Depending on the isoclines, either an extinction of the gene drive males occurs or the gene drive males get extinct and the wild type population persists. The delimitation between both alternatives occurs at a value of 0.5 for FF. In that case, a marginal stability prevails.

This illustrates how important transgene-induced fitness can be at the expense of the outcome of a gene drive application.

Individual-Based Model of an Olive Fly Population with Gene Drive

Individual-based models (IBM) represent activity, interaction and states of single organisms as data structures in a simulation. In most applications, they are spatially explicit and consider environmental structures and temporal variability of environmental states. Therefore, the approach is suitable to analyse the contribution of single aspects and components of the spatial context as well as behaviour and states of organisms to particular results on higher integration levels (e.g. population level), which result from an aggregation of individual activities. In particular, IBM can facilitate a deeper and more detailed understanding of the conditions for specific population dynamic phenomena (Breckling et al. 2006).

Since the possible releases of SPAGE in natural populations pose qualitatively new questions for risk assessment, it was reasonable to demonstrate the utility of the approach also in the context with genetic aspects. Hence, a prototypical IBM structure with regard to the olive fly case study was drafted.

The focus was to prepare a model outline rather than implementing a complete parameterization in any relevant detail. In the current stage of development, the model consists of characteristic features, linking spatial, behavioural and crucial physiological aspects of olive flies. The following features were implemented (Fig. 6.10):

- Spatial structures of the environment, in particular relevant resources like olive tree locations,
- distribution of the individual olive flies as well as aspects of their movement behaviour,
- Temporal characteristics of the development process: Eggs, maggots, pupae and imago,
- Temperature sensitivity of individual development, bottlenecks through low or high temperatures,
- Mating behaviour,
- and we represent a gene drive influence, the distinction of male individuals carrying a gene that leads to the death of females prior to maturity.

The model was written in SIMULA (Dahl et al. 1970; Kirkerud 1989) and compiled with the CIM (SIMULA-to-C) precompiler (https://www.gnu.org/software/cim/). The executable was run on a Lenovo ThinkPad Personal Computer under SUSE Linux. The graphic output was generated using GRPS as an external class.

Since each individual and its specific state is represented, spatial structures as well as the temporal development of the population can be visualized. The simulation was initiated with twelve wild type females, wild type males and gene drive-males each. They were located in a partially overlapping area within direct interaction distance. In the graphic simulation output, green circles mark the location of olive trees, dots show the position of mature and immature life stages. Since random movement processes influence the dispersal process across the simulation area, the results of each model run are unique, however, in terms of probability, typical development trends can be identified. Repeating identical model runs with the only difference in the starting value for the random generator can lead to qualitatively different results of single runs. Random numbers are used to determine co-ordinates of individual movement steps surrounding the current position and mortality events according to temperature dependent conditions. The preliminary simulation results show (Fig. 6.10), that it is important to explore the heterogeneous structure of the response space. Statistical assessment would be a further step to identify result type frequencies in a larger number of model re-runs.

As qualitatively different results of the simulation runs, which were carried out for 630 time steps so far, we observed cases where the gene drive persists while the

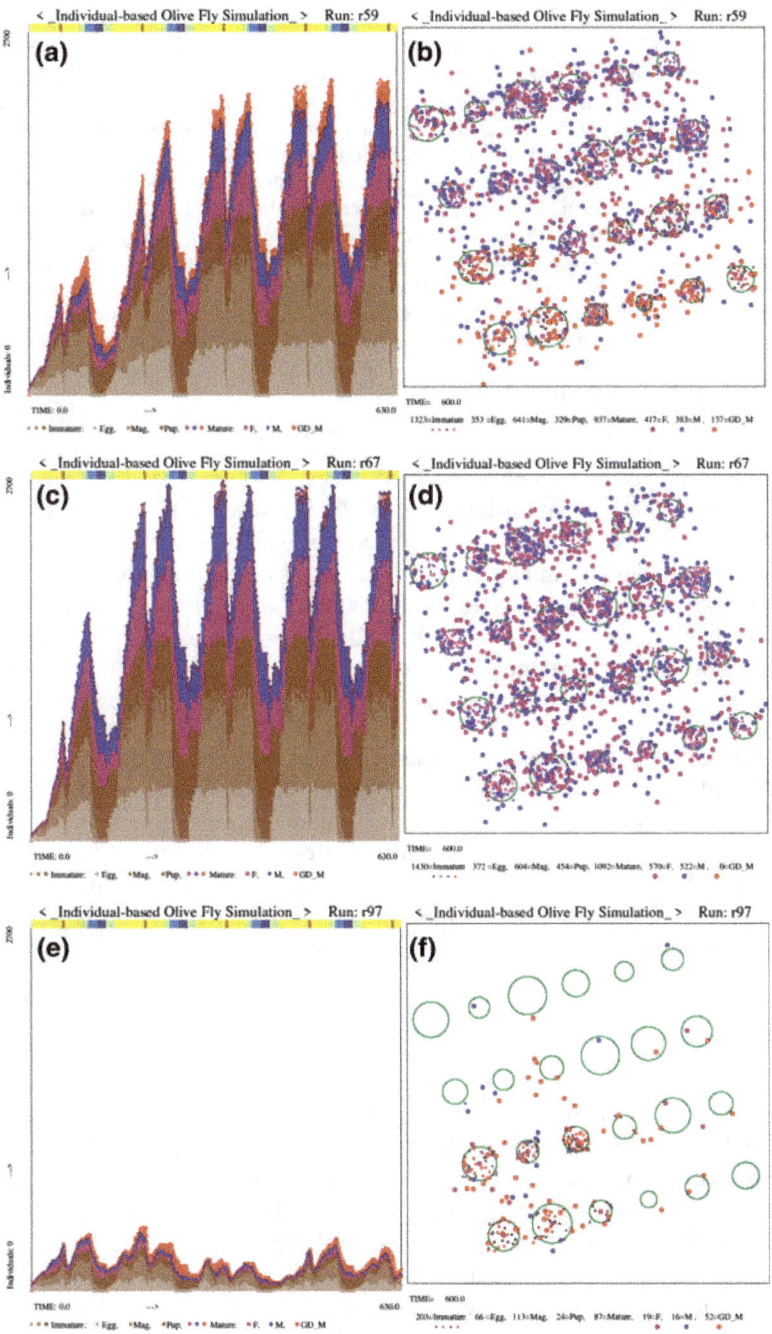
< _Individual-based Olive Fly Simulation_ > Run: r59
(a)
(b)
< _Individual-based Olive Fly Simulation_ > Run: r67
(c)
(d)
< _Individual-based Olive Fly Simulation_ > Run: r97
(e)
(f)

◀**Fig. 6.10** Three different test runs with identical parametrisation but different random seed. Each model run results in a different outcome. **a**, **c** and **e** show the population dynamics over time. Beneath the title line temperature regimes are depicted as a colour bar with blue inducing population decline due to cold conditions and brown marks times of heat decline. **b**, **d** and **f** show the spatial distribution of the population at the final time step. **a** and **b** Simulation run with 59 as start of random number generation: Wild-type and gene drive males persist in a spatially heterogeneous setting. While wild types dispersed, gene drive males largely remain active in the part of the simulated area where they were started. **c** and **d** Simulation run with 67 as start of random number generation: Gene drive males died out, and the wild type population dispersed throughout the simulated area. **e** and **f** Simulation run with 97 as start of random number generation: Gene drive males limited the population expansion and prevented a further dispersal. The population density remains on a low level. The colour code for the individual olive flies is either brown for different immature stages. For the mature stages purple is used for females, blue for males and red for gene drive males (left and right side). Green circles represent trees

population expands throughout the available area. Another simulation run revealed the result, that the population level can be higher when the gene drive carrying individuals died out. Finally, it was observed that under the same parameterization the gene drive males can hinder an expansion across the simulation area and arrest the population density on a comparatively low level. As a preliminary result we can conclude, that there is the potential that under given spatial conditions and under identical physiological specifications, random influences can contribute to a variety of qualitatively distinct outcome in relatively short observation times.

Model studies with the current implementation showed, that the specification is crucially influencing how pronounced the population bottlenecks occur. If phases of exceptionally low or high temperature vary in extent, it would add another significant dimension of uncertainty to the population development.

We suggest to employ individual-based models to analyse in particular how behavioural aspects and spatial heterogeneities contribute to dispersal and persistence processes and how the probability structure of the response space can be understood.

Conclusion

The introduced model concepts give insights how to analyse and understand underlying implications in various aspects of gene drives and population dynamics. It is however important to note, that a model is always a simplification of otherwise highly complex processes. Therefore, a model's applicability to real life is limited. Each model, has its own focal points while even completely dismissing other factor that in reality might have an important influence on the observed process. Although models may facilitate various analyses, they are heavily reliant on what was built into them. Thus, emergent properties of the model were already implied in the model assumptions. Due to this, we did not focus on only one model approach but approached the topic from various sides to allow for a broader view on the vast subject and some of its key aspects.

Nevertheless, models allow insights into complex dynamic processes and furthermore help to discover emergent properties that would have remained undisclosed otherwise. Especially in the context of gene drive releases, which may cause unpredictable consequences, it is important to identify possible hazards and adverse ecological effects on an ecosystem before real-life test trials. To this end, simulations and modelling approaches play a pivotal role.

References

Akbari, O. S., Chen, C.-H., Marshall, J. M., Huang, H., Antoshechkin, I., & Hay, B. A. (2014). Novel synthetic Medea selfish genetic elements drive population replacement in drosophila, and a theoretical exploration of Medea-dependent population suppression. *ACS Synthetic Biology, 3*, 015–928.

Akbari, O. S., Matzen, K. D., Marshall, J. M., Huang, H., Ward, C. M., & Hay, B. A. (2013). A synthetic gene drive system for local, reversible modification and suppression of insect populations. *Current Biology, 23*, 671–677. https://doi.org/10.1016/j.cub.2013.02.059.

Bocaccio, L., & Petacchi, R. (2009). Landscape effects on the complex of *Bactrocera oleae* parasitoids and implications for conservation biological control. *BioControl, 54*, 607–616.

Breckling, B., Middelhoff, U., & Reuter, H. (2006). Individual-based models as tools for ecological theory and application: Understanding the emergence of organisational properties in ecological systems. *Ecological Modelling, 194*, 102–113. https://doi.org/10.1016/j.ecolmodel.2005.10.005.

Buchman, A., Marshall, J. M., Ostrovski, D., Yang, T., & Akbari, O. S. (2018). Synthetically engineered Medea gene drive system in the worldwide crop pest *Drosophila suzukii*. *PNAS, 115*, 4725–4730.

Burt, A., Coulibaly, M., Crisanti, A., Diabate, A., & Kayondo, J. K. (2018). Gene drive to reduce malaria transmission in Sub-Saharan Africa. *Journal of Responsible Innovation, 5*, S66–S80. https://doi.org/10.1080/23299460.2017.1419410.

Comins, H. N., & Fletcher, B. S. (1988). Simulation of fruit fly population dynamics, with particular reference to the olive fruit fly, *Dacus oleae*. *Ecological Modelling, 40*, 213–231.

Daane, K. M., & Johnson, M. W. (2010). Olive fruit fly: Managing an ancient pest in modern times. *Annual Review of Entomology, 55*, 151–169. https://doi.org/10.1146/annurev.ento.54.110807.090553.

Dahl, O.-J., Myhrhaug, B., & Nygaard, K. (1970). *SIMULA common base*. Norway: Norwegian Computer Center.

Dhole, S., Vella, M. R., Lloyd, A. L., & Gould, F. (2018). Invasion and migration of spatially self-limiting gene drives: A comparative analysis. *Evolutionary Applications, 11*, 794–808. https://doi.org/10.1111/eva.12583.

DiCarlo, J. E., Chavez, A., Dietz, S. L., Esvelt, K. M., & Church, G. M. (2015). RNA-guided gene drives can efficiently bias inheritance in wild yeast. *Nature Biotechnology, 33*, 1250–1255. https://doi.org/10.1038/nbt.3412.

Esvelt, K. M., Smidler, A. L., Catteruccia, F., & Church, G. M. (2014). Concerning RNA-guided gene drives for the alteration of wild populations. *eLife, 3*. https://doi.org/10.7554/eLife.03401.

Fischer, L. (2018). Wer die Mücken auslöscht, ist auch Malaria los. Zeit Online - Wissen, 24.9. https://www.zeit.de/wissen/gesundheit/2018-09/crispr-methode-malaria-muecken-bekaempfung-studie.

Fu, G., Lees, R. S., Nimmo, D., Aw, D., Jin, L., Gray, P., et al. (2010). Female-specific flightless phenotype for mosquito control. *PNAS, 107*, 4550–4554. https://doi.org/10.1073/pnas.1000251107.

Galizi, R., Doyle, L. A., Menichelli, M., Bernardini, F., Deredec, A., Burt, A., et al. (2014). A synthetic sex ratio distortion system for the control of the human Malaria mosquito. *Nature Communications, 5*, 3977. https://doi.org/1038/ncomms4977.

Genç, H., & Nation, J. L. (2008). Maintaining *Bactrocera oleae* (Gmelin.) (Diptera: Tephritidae) colony on its natural host in the laboratory. *Journal of Pest Science, 81*, 167–174. https://doi.org/10.1007/s10340-008-0203-3.

Gould, F., Huang, Y., Legros, M., & Lloyd, A. L. (2008). A killer-rescue system for self-limiting gene drive of anti-pathogen constructs. *Proceedings of the Royal Society B: Biological Sciences, 275*, 2823–2829. https://doi.org/10.1098/rspb.2008.0846.

Groß, J. (2016). Turbo für die Evolution. Süddeutsche Zeitung. February 19. https://www.sueddeutsche.de/wissen/eingriff-ins-erbgut-die-gene-die-ich-rief-1.2867148.

Hammond, A., Galizi, R., Kyrou, K., Simoni, A., Sinicalchi, C., Katsanos, D., et al. (2016). A CRISPR-Cas9 gene drive system targeting female reproduction in the malaria mosquito vector Anopheles gambiae. *Nature Biotechnology, 34*, 78–85. https://doi.org/10.1038/nbt.3439.

Jacobs, A. (2015). Eine Kettenreaktion gegen Malaria. Neue Bürcher Zeitung. June 16. https://www.nzz.ch/wissenschaft/medizin/eine-kettenreaktion-gegen-malaria-1.18563485.

Kirkerud, B. (1989). *Object-oriented programming with SIMULA. International Computer Science Series*. Wokingham, England; Reading, Mass: Addison-Wesley Pub. Co.

Knipling, E. F. (1955). Possibilities of insect control or eradication through the use of sexually sterile males. *Journal of Economic Entomology, 48*, 459–462. https://doi.org/10.1093/jee/48.4.459.

Kokkari, A. I., Pliakou, O. D., Floros, G. D., Kouloussis, N. A., & Koveos, D. S. (2017). Effect of fruit volatiles and light intensity on the reproduction of *Bactrocera* (Dacus) *oleae*. *Journal of Applied Entomology, 141*. https://doi.org/10.1111/jen.12389.

Montiel Bueno, A., & Jones, O. (2002). Alternative methods for controlling the olive fly, *Bactrocera oleae*, involving semiochemicals. *International Organization for Biological and Integrated Control of Noxious Animals and Plants West Palaearctic Regional Section (IOBC/WPRS) Bulletin, 25*, 1–11.

Nardi, F., Carapelli, A., Dallai, R., & Roderick, G. K. (2005). Population structure and colonization history of the olive fly, *Bactrocera oleae* (Diptera, Tephritidae). *Molecular Ecology, 14*, 2729–2738. https://doi.org/10.1111/j.1365-294X.2005.02610.x.

Pontikakos, C. M., Tsiligiridis, T. A., & Drougka, M. E. (2010). Location-aware system for olive fruit fly spray control. *Computers and Electronics in Agriculture, 70*, 355–368.

Reeves, R. G., Bryk, J., Altrock, P. M., Denton, J. A., & Reed, F. A. (2014). First steps towards underdominant genetic transformation of insect populations. *PLoS ONE, 9*(e97557), 1–9. https://doi.org/10.1371/journal.pone.0097557.

Sánchez-Ramos, I., Fernández, C. E., González-Núnez, M., & Pascual, S. (2013). Laboratory tests of insect growth regulators as bait sprays for the control of the olive fruit fly, *Bactrocera oleae* (Diptera: Tephritidae). *Pest Management Science, 69*, 520–526. https://doi.org/10.1002/ps.3403.

Sharaf, N. S. (1980). Life history of the olive fruit fly, *Dacus oleae* Gmel. (Diptera: Tephritidae), and its damage to olive fruits in Tripolitania. *Journal of Applied Entomology, 89*, 390–400. https://doi.org/10.1111/j.1439-0418.1980.tb03480.x.

Thomas, D. D., Donnelly, C. A., Wood, R. J., & Alphey, L. (2000). Insect population control using a dominant, repressible, lethal genetic system. *Science, 287*, 2474–2476.

Tsiropoulos, G. J. (1972). Storage temperatures of eggs and pupae of the olive fruit fly. *Journal of Economic Entomology, 65*, 100–102. https://doi.org/10.1093/jee/65.1.100.

Tsitsipis, J. A. (1980). Effect of constant temperatures on larval and pupal development of olive fruit flies reared on artificial diet. *Environmental Entomology, 9*, 764–768.

Voulgaris, S., Stefanidakis, M., Floros, A., & Avlonitis, M. (2013). Stochastic modeling and simulation of olive fruit fly outbreaks. *Procedia Technology, 8*, 580–586.

Ward, C. M., Su, J. T., Huang, Y., Lloyd, A. L., Gould, F., & Hay, B. A. (2011). Medea selfish genetic elements as tools for altering traits of wild populations: A theoretical analysis. *Evolution, 65*, 1149–1162. https://doi.org/10.1111/j.1558-5646.2010.01186.x.

7

Gene Drives and the Mitigation of Risk

Bernd Giese, Arnim von Gleich and Johannes L. Frieß

Introduction

With the release of organisms carrying a gene drive (GD) a fundamental change in the release practice of GMOs will take place. GMOs as GD-carrying organisms (GDO) will then arise from wild populations in the field, not in controlled numbers in the laboratory or a breeding facility (Simon et al. 2018). The inherent ability of GDs to overcome the limits of Mendelian inheritance even for traits with detrimental effects on their fitness make them an ideal tool for the efficient manipulation of wild populations of sexually reproducing species. The idea of using GDs was originally inspired by the discovery of naturally occurring mechanisms like transposable elements or meiotic drives that trigger a super-Mendelian dissemination of traits. After early proposals to harness chromosomal translocations for population control (cp. Curtis 1968) the use of selfish genetic elements was proposed in 1994 by Hastings. The idea of using homing endonucleases to build self-replicating drives which are now realized with the help of the versatile molecular scissor CRISPR-Cas9 was already put forward in 2003 by Austin Burt (Burt 2003). Besides the spread of new traits with so called 'conversion drives', 'suppression drives' are aiming at a reduction or even a regional extinction of pest species or vectors of pathogens. Amongst other application fields suppression drives are envisaged for combating malaria by strongly reducing the number of some mosquito species which are the prime vectors for infectious diseases like malaria and dengue (Macias et al. 2017). Anticipated as a highly specific replacement for pesticides, GDs are considered to be applied against a number of invasive species that have become agricultural pests like the

B. Giese (✉) · J. L. Frieß
Institute of Safety/Security and Risk Sciences, University of Natural Resources and Life Sciences, Vienna (BOKU), Austria
e-mail: bernd.giese@boku.ac.at

A. von Gleich
Department of Technological Design and Development, Faculty Production Engineering, University of Bremen, Bremen, Germany

cherry fruit fly *Drosophila suzukii* in California (Buchman et al. 2018) or rodents like mice or rats in New Zealand which pose a serious threat to agriculture and the native environment (Dearden et al. 2018). Even weeds have been proposed as targets for suppression drives (Neve 2018). Conversion drives are currently developed to render mosquitoes immune to pathogens or inhibit the penetration of ripening fruits by the cherry fruit fly due to a non-functional ovipositor (Regalado 2017).

A number of reasons for concern have been raised in the course of the discussion about GD development and their potential applications (Esvelt and Gemmell 2017; Ledford 2016; National Academies of Sciences 2016; Oye et al. 2014). They represent a potentially powerful genetic tool with unprecedented range in time and space. Current releases of GMOs are limited to a certain number of engineered organisms (mostly plants), a limited timeframe and a limited area with more or less established separation from wild relatives to prohibit vertical gene transfer to subsequent generations. With GDs a fast vertical gene transfer becomes the aim of the GMO-application. To ensure control of GDs or even strive for a kind of functional reversibility, a number of options have been proposed in recent years. A review of their potential, their reliability and their developmental stage is still missing. Furthermore, alternative approaches are excluded so far in a comparative evaluation. The present work should help to close this gap.

In addition to technical variations of GDs, which may yield an improved reliability or decrease their hazard and exposure potential as sources of possible risks, alternatives to GDs are as well included in the assessment which is given in this chapter. But the technology itself is only one factor in the generation of risks.

Additional exposure- and hazard potentials depend on the qualities and the vulnerability of the ecosystems into which the gene drives are introduced and on the specific aims and contexts of the gene drive application (e.g., agriculture, disease control or nature conservation) if the GD-system is not per se containing toxic or allergenic substances. Beyond these known adverse properties nearly any biochemical quality, e.g. an enzymatic feature, may turn out hazardous in a particular context. Thus, corresponding non-knowledge on the final behaviour hinders a characterization of the total hazard potential in very early innovation phases where results of experimental tests and specific application conditions are not available. Thus, as a precautionary approach especially in anticipation of environmental release, it is advisable to focus more on the exposure potential instead of the hazard potential. A high exposure strongly increases the possibility of unforeseen interactions in the environment, and thus increases the realm of ignorance about possible adverse effects. That is the lesson learned from the release of persistent synthetic chemicals into the environment (e.g. Chlorofluorocarbons [CFC] and Persistent Organic Pollutants [POPs]). To focus on exposure relevant qualities yields options on how to limit or even decrease the exposure potential emanating from GDs. It may thereby reduce the potential for unforeseen and unmanageable interactions of GDs in the environment. Reducing the exposure potential is thus a promising approach of risk reduction for GD.

The exposure potential of GDs is determined by qualities of the GD or the GDO that are related to (a) the spatial and (b) the temporal spread. These could be for example

- stability of the GD against inactivation by mutations,
- impact of the GD on the fitness of the target species,
- frequency of inheritance

as GD-specific qualities. With regard to the target species the following qualities may have an influence:

- mobility,
- life expectancy,
- inheritance,
- number of offspring,
- probability of crossbreeding,
- frequency of releases and initial number of released individuals carrying the GD,
- regional distribution of the target population,
- interconnections between subpopulations.

In case of gene drives that are exclusively applied in laboratories, exposure to the environment is mainly determined by the containment of the experimental settings. We may call this an *extrinsic containment*. For an overview of extrinsic containment strategies see (Akbari et al. 2015; Benedict et al. 2008). If barriers are characteristic for extrinsic containment, ecological containment can be seen as a special form of this type of containment where spatial separation serves as a means for safety. Here, wild type populations of the target species or wild relatives are lacking in the geographic region where the GDOs are released or where the GD experiments are performed in a laboratory. Additionally, environmental conditions should not favour the settlement of these wild species. With regard to the safety of laboratory experiments with GDO, ecological containment is supported by environmental conditions that are not favourable for the respective GDO-species, e.g. regarding temperature. However, ecological containment is an option of limited reliability because GDOs might be transported intentionally or unintentionally along with other cargo in ships, cars or planes and some may as well survive unfavourable climate (Min et al. 2017b).

Additionally, laboratory safety can be improved by an appropriate *intrinsic containment*.[1] That is GDs or specific target species are dependent on synthetic substances or limited in spread due to their specific technical organization. Because extrinsic containment is practically only an option for GDs applied in laboratories, in this chapter we will focus on approaches for intrinsic containment with relevance for GDO that are to be released into the environment.

[1]Cf. Wright et al. (2013, 1223): "Biology can achieve a lot in a contained environment; however, physical containment alone offers no guarantees. For example, no matter how ingenious a protective device or material may be for a GMM field application, an inventive way will eventually be found by an operator to compromise it. Failure in this case is a matter of when, not if. Although some form of physical containment is obviously prudent, inbuilt biological mechanisms remain crucial to biosafety."

Intrinsic Containment

The intrinsic containment of a GDO is either due to the reproductive incompatibility of the target species with wild type strains and related species or caused by the specific character of the GD. For instance, in case of homing endonuclease-based GDs (HEG-drives) the latter may be due to a unique target sequence. Accordingly, Min et al. differentiate between reproductive and molecular confinement as variants of intrinsic containment (Min et al. 2017b, p. 55).

For GDOs used *in the laboratory,* a number of special options for intrinsic containment are available that have been already used as safety measures for experiments with GMOs. For applications in laboratory facilities it is advisable to use organisms that are not viable outside laboratory conditions. Containment strategies can make use of the following options:

- dependency on the supply of a synthetic substance which is only available in the laboratory, (see RIDL, Chap. 2),
- a kill switch which is activated when a certain food compound is lacking,
- reproductive containment using laboratory strains unable to produce viable offspring with wild conspecifics, e.g., the use of Drosophila with compound autosomes, where the left arms of two chromosomes are joint together in one chromosome and the right arms in another, making these specimen infertile with wild types (Akbari et al. 2015).

For GDO-species that are determined to be released into the environment, safety strategies become more challenging because it is intended that gene drives spread within a population by mating of GDOs with their wild conspecifics. Meanwhile a number of approaches to limit the spread of GDOs in time and space have been proposed (Esvelt et al. 2014; Noble et al. 2019). Within the following section potential options are presented.

Safety Options for GDO-Releases

Safety strategies for GD applications can be grouped into techniques that represent either gene drive modifications and other transgenic constructs respectively or rather alternative approaches which are based on naturally occurring mutations and parasitic infections that enable population control in a comparable way. Options for both types of approaches, either relying on genetic engineering or harnessing naturally occurring anomalies are presented in the following chapters.

Molecular Modifications of Gene Drives as Safety Strategy

Split Drive

The idea of a split drive to limit the uncontrolled spread of a GD is based upon the separation of the genetic components of CRISPR/Cas based HEG-drives. To that end, the endonuclease gene and the genetic information of the single guide RNA (sgRNA) can be located at different loci in the genome of which only one of the genes is inherited as a GD. For example, if the sgRNA code for a target sequence resembles its own insertion site, only the inheritance of the sgRNA will be biased in a super-Mendelian fashion. Inheritance of the endonuclease gene is by contrast determined by Mendelian dynamics and should therefore lead to a loss of the Cas9 gene and thus limit the spread of the sgRNA gene after a few generations as long as Cas9 does not provide a fitness gain (cf. DiCarlo et al. 2015).

Malfunction of a split drive can be caused by molecular recombination events of the genome that may move the Cas9-gene adjacent to the sgRNA sequence. If reading frames are intact the result would be a complete and therefore potentially autonomous GD consisting of the information for the endonuclease as well as a sgRNA. At least the non-intended integration of sgRNA-sequences has already been observed (Li et al. 2016). Homology directed repair of the next sgRNA guided cleavage of a target site would then result in copying of sgRNA and endonuclease genes. However, the probability for such an event is low and it can be further reduced by a low homology between the locations of both elements of the split drive within the genome (Akbari et al. 2015). Additionally, developers of GD recommend to combine this strategy with a second form of containment (Akbari et al. 2015).

Besides a separation within the genome, other variants of split drives are imaginable. At least for some eukaryotic species, the genetic information of the endonuclease Cas9 can be located episomally, outside the genome on an extrachromosomal plasmid. DiCarlo et al. experimentally verified the function of a split gene drive with episomal Cas9 gene in yeast (DiCarlo et al. 2015). But in addition to a verification of the gene drive-biased inheritance an assessment of the limiting effect of this kind of split drive system is still lacking. The bias of inheritance ceases with each generation as plasmids get lost. And even if the endonuclease gene is moved inside the genome by recombination which is a rare but not impossible event, this gene will most likely be inherited by Mendelian dynamics and therefore the "drive" of the sgRNA fades over the next generations.

An even more expanded version of a split drive would be a constellation with different strains carrying parts of the genetic information of a gene drive. For example, the gene for Cas9 can be part of the genome of a strain that mates with a sgRNA-bearing strain. In such an approach the strain carrying Cas9 has to be released continuously to keep the drive running, because the Cas9-gene is only passed to offspring with a 50% chance of inheritance if the sgRNA targets its own insertion sequence (Akbari et al. 2015).

Daisy Chain Drives

In a daisy drive-system, a number of gene drives is dependent on each other in a linear (or circular) manner, in that each of the drives' sgRNAs is encoded for a target site which consists of the flanking elements of the next drive in the daisy chain. Therefore, no element of the chain drives itself. The single drives of a chain can even be located on different chromosomes. In a linear chain (of at least two elements), the first drive has no predecessor which would cut a target site in which the first drive could be integrated by homologous recombination. Therefore, it will be the first element of the daisy chain that gets lost by the means of natural selection. Accordingly, the other elements of the chain will successively get lost over time. In the daisy drive proposed by Noble et al. (2019), the last element of the chain carries the "payload" (the cargo gene). If finally the last sgRNA of the chain is lost, the top element will fade away as well, if it does not deliver any fitness gain (Noble et al. 2019). As for split drives, recombination may create an independent drive which then overcomes the limiting effect of the daisy chain.

In 2017, Min et al. proposed a "Daisy field" drive-system (Min et al. 2017b), where multiple sgRNAs are encoded separately from the locus harbouring CRISPR nuclease and a potential payload gene, but all sgRNAs share the same target sequence. Compared with a daisy chain drive, the daisy field system works with just a single cut-and-copy event and thus should be more reliable and less prone to non-intended recombination events that may create a global drive. According to Min et al. the fitness cost should be small, because the multiple elements of a daisy field drive (except for the cargo genes) consist of sgRNAs. Only for a number of initial generations, the genetic information for the nuclease and the payload is inherited to all offspring (by homologous recombination) due to the fact that with each generation the sgRNA daisy elements (N_{sgRNA}) are inherited with only 50% chance. Because with every generation the average number of sgRNAs per organism is cut in half, the nuclease and cargo-genes will be inherited by the drive for roughly ($N_{sgRNA} + 1$) generations (Min et al. 2017b). According to this theory, the initial number of sgRNAs should therefore be a means to tune the spread of a drive. Daisy field drives can be combined with a daisy drive chain for instance as the first element of a drive chain (Min et al. 2017b).

In order to prevent the accidental generation of a global drive by recombination events that would move gRNA adjacent to the nuclease, sequences of gRNA and nuclease (including the cargo genes) should not have sequence homology. Min et al. suggest to have not more than 12 base pairs homology. Additionally, they recommend to place the nuclease more than 100 kb apart from gRNA repeat sequences (Min et al. 2018).

In a further prepublication, Min et al. propose the concept of "Daisy quorum" drives as an extension of a daisy drive-application by the subsequent release of wild-type organisms or a suppression drive targeting the previously altered population. This combination should lead to a low frequency of the engineered genes which then theoretically get lost over time by natural selection, if it does not provide a fitness gain for the organism expressing them (Min et al. 2017a).

Limitation by Secondary Releases

Secondary releases of sexually compatible members of the target species have been mentioned quite early as a method to limit the spread of GDs and even as a means to reverse the functionality of the drive in the already affected individuals. The proposed approaches range from releases of sterile wild type individuals that should breed with the genetically altered organisms, thereby slowing down the spread of the drive (Montell cited in McFarling 2017), to the release of GDOs equipped with overwriting drives that target the initial drive sequence. Particularly tricky approaches for the removal of CRISPR/Cas9-drives should even function without a full GD-functionality: They only rely on gRNAs whose target sequences are flanking the sequence of the previously released GD or are located within the coding sequence of Cas9. Due to the cellular presence of Cas9 from the released drive, which is now guided by the gRNA of the removal construct, excision or disruption of the GD and replacement with the coding cassette of the removal construct is enabled (cp. Zentner and Wade 2017).

Most probably, all these types of approaches for secondary releases are rather imperfect in their reliability as a means to limit or reverse the impact of released GDs because their spread must at least cover the spatial and numerical distribution of the initial drive. In particular, with regard to overwriting drives, a second (overwriting) drive has to reach every individual that was altered by the initial drive to exclude the possibility of recurrence—which cannot be excluded at least for very effective drives with a low threshold such as CRISPR/Cas-based systems. Furthermore, as long as the GD is not strictly threshold-dependent, a release of sterile wild types is probably only able to slow down the spread of the drive. However, overwriting drives have been discussed in the community of scientists engaged in GD-development and Esvelt, on his webpage, demands that an overwriting drive should be built in parallel to any new gene drive.[2] According to Esvelt an overwriting—or "immunizing reversal" drive as he calls it—should not only target the individuals that are already altered by the initial drive: Besides overwriting the GD-code in the latter, it should render the wild type-population immune to further spread of the initial drive. He admits that "reversal" only refers to the phenotype, not the genotype of the altered organisms, because the second drive will not be able to restore the original genetic code. Traces of the genetic information of Cas9 and the sgRNA will remain in their genome.

Limitation by Dependence

Besides a specific genetic structure that may serve as a means to limit the spread of GDs, their ongoing super-Mendelian inheritance could be limited by different types of dependence. External factors that may have an impact on the dynamics of GD

---

[2] cf. https://www.sculptingevolution.org/genedrives/safeguards.

distribution are environmental conditions, a specific (synthetic) target sequence or a (synthetic) inductor molecule.

In the latter case, the inductor is necessary to induce the expression of the endonuclease or the sgRNA (if a CRISPR/Cas-based GD is used). If the toxin in toxin-antidote drives is constitutively repressed, an inductor would be necessary to release the toxin and thereby activate the drive. But this method may turn out as difficult to realize for multicellular eukaryotic organisms, because the inductor has to be present in the germline and therefore cross several barriers of the organism's body. As an opposite strategy to an inductor, a toxin might be used which only impacts GDOs due to a sensitivity mediated by the genomic manipulation or the cargo of the gene drive.

If a HEG-drive is engineered to target a specific sequence that is unique to a certain number of individuals, it can be used to limit the spread of the drive to subpopulations of a species or previously released GMOs. Esvelt and colleagues called the limitation to subpopulations a "precision drive" (cp. Esvelt and Gemmell 2017; Min et al. 2017b, p. 49). But according to Esvelt et al. it could be difficult to realize this drive type. First, to assure that the drive targets at least the subpopulation, it has to withstand the occurrence of resistant alleles. For that purpose, they suggest to have a multiplex drive with at least three target sites. Additionally, these sites have to be located within the sequence of essential genes. Moreover, for the application of CRISPR/Cas drives, these "natural" sequences have to contain a protospacer adjacent motif (PAM) to enable binding of the endonuclease to the target site.

These obstacles could be overcome if the GD targets only synthetic sequences encoded in genetically engineered organisms. As "synthetic site targeting" this safety approach was tested in yeast in an initial experiment (DiCarlo et al. 2015). A major advantage of this approach is the fact that depending on the sequence similarity with natural sequences the sgRNA of an HEG-drive must undergo several mutations before it may serve to place a drive in a natural sequence. For the application in isolated populations e.g., on islands, Min et al. suggested to use target sequences for HEG-drives that are recoded by an initial drive to provide an appropriately prepared population (Min et al. 2017b, p. 49).

The idea of Craig Montell is an example for a dependency on an environmental factor: He suggested to engineer mosquitoes with a self-destruction mechanism which is activated when an environmental parameter, e.g. temperature, reaches a threshold (Montell cited in McFarling 2017). In order to catch every gene drive mosquito by this technique, it has to be included in the cargo of the drive. Besides the necessary increase in size for the additional cargo information, the major drawback of this approach most probably lies in its vulnerability to mutations in the self-destruction mechanism.

Limitation by Genetic Instability

Experimental tests of CRISPR/Cas-GDs revealed a significant restraining impact of resistance alleles in target populations. Selection of resistance to a CRISPR/Cas-GD was first documented by Hammond et al. in 2017. After an initial increase of GDO in caged mosquito populations over less than 10 generations they observed a gradual increase of the ratio of resistant alleles within the experimental time frame of 25 generations (Hammond et al. 2017). Resistant alleles may occur due to Non-Homologous End Joining (NHEJ), Microhomology Mediated End Joining (MMEJ), by incomplete Homology Directed Repair (HDR) or may originate from sequence variations within the population (Champer et al. 2017). Within the sequence of essential genes of a species the probability is high that mutations compromise the viability of the organism. Thus target sites within essential genes that are highly conserved among the members of a species are likely to confer more stability with regard to the spread of the GD. As Kyrou et al. have shown, this strategy is successful in suppressing the selection of resistant alleles in *Anopheles gambiae* (Kyrou et al. 2018). An approach for GD-limitation—at least as an additional feature—could be established by a high probability for mutations due to only a single target site within a non-essential gene and only a single sgRNA locus.

Alternative Approaches to Synthetic Gene Drives

Population control can potentially be achieved with alternative approaches as well. Alternative approaches can be divided into options based on genetic engineering as the use of transgenes, mutations causing sterility and the application of naturally occurring effects with a dampening impact on population growth. Important examples of both groups are presented within the next two chapters.

Release of Insects Carrying a Dominant Lethal Allele (RIDL) as an Alternative Genetic Engineering Approach

For the RIDL-approach laboratory-reared organisms equipped with a dominant lethal gene are mass released to reduce the number of offspring in a wild population (Thomas et al. 2000). The dominant lethal gene prohibiting development of the offspring is cloned into the insects. Offspring dies at zygotic, larval or pupal stage. There are two varieties of RIDL. In the bi-sex RIDL approach, the offspring of both sexes die in premature stages (Harris et al. 2011; Phuc et al. 2007). In the female specific RIDL approach (fsRIDL), only female offspring die (Schliekelman and Gould 2000; Thomas et al. 2000). Heterozygous sons then pass the lethal gene to half of both sexes of their offspring, of which the inheriting females will die. Female specific RIDL strains have been developed for *Aedes aegypti* and *Aedes albopictus*, using flightlessness as a lethal trait (Alphey et al. 2013). Since only genetically engineered males are released, the bi-sex RIDL approach is considered self-limiting, while fsRIDL has to be considered self-sustaining, albeit for a limited number of generations.

RIDL organisms are furthermore equipped with a fluorescent marker for distinction from wild type animals and a bistable switch, a tetracycline-repressible transcription activator (tTAV) which binds to the promotor of the lethal gene. In the presence of tetracycline, which is included in the laboratory organisms' diet, the transcription activator is repressed, allowing for normal procreation (Thomas et al. 2000).

However, apart from the observation that the lethality of RIDL-offspring fails in 3% of cases there is evidence that due to "naturally occurring" contamination of food with tetracycline, RIDL-offspring become viable and survive (Rodriguez-Beltran 2012). Another drawback with regard to the safety of this approach is that mechanical sex sorting fails in 0.33% of cases (Lacroix et al. 2012).

In caged trials it was shown that RIDL mosquitoes have an estimated competitiveness to wild type males of 0.56 (Harris et al. 2011). Male RIDL-olive flies are outcompeted for mating in cages, with 46% of females mating RIDL-flies (Ant et al. 2012). Thus, to continually suppress a population, a high number of periodical releases with large numbers of insects is required. Field trials with mosquitoes were carried out in the Cayman Islands, Malaysia, Brazil, and Panama (Alphey 2014; Gorman et al. 2015; Subbaraman 2011).

Resistance by Transgenes

Different approaches have been published so far for the suppression of mosquito populations e.g., by the expression of antibodies against the malaria parasite or RNAi-expression to suppress arbovirus replication, but expression of transgenes may represent a fitness cost and therefore might get lost before significant results in population control are achieved, moreover resistance of pathogens may evolve over time (Alphey et al. 2013).

Population Control by Mutagenesis: The Sterile Insect Technique (SIT)

SIT is applied for population suppression of disease transmitting mosquitoes or pest insects like the Mediterranean fruit fly since the middle of the 20th century (Dyck et al. 2005). The technical approach of SIT consists in the release of masses of sterile insects. Sterile males are the preferred candidates for release because release of both sexes (a) seems to weaken the suppressing effect due to mating between the sterilized males and females and (b) in case of mosquito control a release of females would increase the potential of biting humans. Mating of the released males with wild females leads to a decline of the target population and in extreme cases to its collapse. SIT-insects are sterilized by radiation or chemicals. The induction of dominant lethal mutations in the treated sperm leads to death of the majority of eggs that are fertilized by this sperm. For SIT it is important that sterile males are not agametic. Spermatozoa still have to be produced to keep up sperm competition with sperm from fertile males in remating females (Alphey et al. 2013).

Harnessing Naturally Occurring Phenomena for Population Control

The Trojan-Female-Technique (TFT)

For some species male fertility seems to depend on mutations in mitochondrial DNA. These male-specific deleterious mutations escape selection processes in the female germline, because mtDNA is maternally inherited (Beekman et al. 2014; Frank and Hurst 1996 cited in Wolff et al. 2017). Their application may yield a long lasting approach for population control across several generations. A corresponding approach is called Trojan Female Technique (Gemmell et al. 2013). By using TFT for population control, genome editing can be avoided if female insects with mtDNA-mutations are selected that are highly effective with regard to male infertility. Therefore, neither transgenic nor genome edited organisms are necessary for a TFT-approach and legal preparation of application would be clearly simplified.

mtDNA-mutations causing sub- or complete infertility have been reported for Drosophila, seed beetles, hares and humans (cf. Wolff et al. 2017). A first proof of concept in fruit flies was shown by Wolff et al. (2017). They used a mitochondrial haplotype from a population of *D. melanogaster* that leads to complete male sterility in combination with a particular genomic background in the nucleus. The suppressive effect persisted over 10 generations in a density controlled population were egg numbers in each generation are carefully regulated. Over the course of the experiment (10 generations) no mutations in the nuclear genome could be detected that compensated for mutations in the mitochondrial DNA which caused the male infertility. Additionally, no reduction in frequency of the TFT-haplotype was detected. This observation highlights the power of the TFT approach as it may be sufficient to apply it in a single release and achieve a suppression over multiple generations. However, the effect was smaller than expected and unfortunately for more natural conditions without density control (stochastic contractions and expansions in population size are possible) no significant effect could be detected (Wolff et al. 2017).

Wolbachia Parasites and Cytoplasmic Incompatibility

In 1967 it was first described that the bacterial endosymbiont *Wolbachia pipientis* is able to bias inheritance of infected mosquitoes leading to the spread of the bacteria in populations of the host (cf. Macias et al. 2017, p. 4). Due to its cytoplasmic localization Rickettsia bacteria of the genus *Wolbachia* are maternally passed on to their offspring and infect ~52% of the terrestrial arthropod species (Weinert et al. 2015, p. 3). As determined in *Drosophila* and *Aedes aegyti*, infection with *Wolbachia* bacteria may reduce the host life span by roughly half (Lin et al. 2012). Furthermore, *Wolbachia* poses effects such as male killing, feminization, parthenogenesis and cytoplasmic incompatibility (Burt 2014). In terms of population control or measures against infectious diseases two approaches are possible with *Wolbachia*:

1. An infection of mosquitoes with certain strains of *Wolbachia* can be used to reduce susceptibility of the insects to a range of different pathogens (Alphey 2014; Alphey et al. 2013; Carrington et al. 2018). The *Wolbachia* strain *w*Mel for example blocks the development of dengue in *Ae. aegypti.*
2. If male mosquitoes are infected with specific *Wolbachia* strains, mating with uninfected females or females infected with different *Wolbachia* strains is unsuccessful due to a cytoplasmic incompatibility of eggs and sperm (wild female mosquitoes are usually infected with other *Wolbachia* strains) (Blagrove et al. 2012; Burt 2014).

The latter corresponds to a population suppression approach and is called "incompatible insect technique (IIT)" (Burt 2014). Infected females will always produce infected offspring. The application is complicated because for IIT with infected males the release of only male carriers is absolutely essential as a single female carrier could potentially drive the alien *Wolbachia* strain into the wild population, mitigating the desired effect. The latter effect would be desired in an invasive application, which is potentially self-sustaining. In another invasive application, organisms infected with two different *Wolbachia* strains, each incompatible with the other, are released. This is the so called bidirectional approach. The result would be the formation of three different subpopulations, wild type, *Wolbachia* strain A and *Wolbachia* strain B.

The use of *Wolbachia* bacteria for population and disease control may come along with a number of drawbacks. As reported in Alphey et al. (2013) infection with *Wolbachia* could cause a selection towards viruses with higher titre in human-biting mosquitoes (Alphey et al. 2013). Furthermore, due to the need for human blood to produce viable eggs, wMel-infected mosquitoes may develop an increased preference to bite humans (Alphey et al. 2013). Evolutionary changes might affect the relationship between the parasite and its host: Resistance of mosquitoes against viruses like dengue as a result of their immune response against *Wolbachia* might be dampened due to a co-evolution of the mosquito and the *Wolbachia* parasite and even the virus may adapt over time (Macias et al. 2017, p. 13). For *Wolbachia*-based approaches experience was gained by a number of field trials in recent years (see Table 7.1).

Overview of Potential Safety Mechanisms

The different strategies presented here that may help to overcome the potential risks of GD vary remarkably in that they on the one hand rely on genetic engineering—partially even consist in GD-variants—and on the other hand they represent real alternatives which are not based on genetic engineering at all. The sheer variety

Table 7.1 Known *Wolbachia* field trials (updated collection of Alphey 2014)

Date	Location	Method	Outcome
2010	French Polynesia	IIT	Sustained release of *Aedes polynesiensis* males infected with a *Wolbachia* strain from *Aedes riversi* induced sterility in a target population
2011–Present	Australia	Invasive *Wolbachia*	Release of infected male and female *Aedes aegypti* led to the invasion and establishment of different *Wolbachia* strains; releases in multiple additional areas are in progress
2013–2018	Vietnam	Invasive *Wolbachia*	Release of wMelPop-infected male and female *Aedes aegypti* on an island
2017	California	IIT	Release of male *Wolbachia*-infected *Aedes aegypti* to suppress the natural population in Fresno County

shows that at least theoretically there are options for population control besides the application of GDs. Nonetheless, these approaches differ strongly in their qualities with regard to the aim of reducing exposure to GD and minimizing potential hazards associated with their release. And besides the fact that the effectivity of most of these options is not yet experimentally verified, they are connected to different vulnerabilities that may preclude particular applications. To facilitate an overview on the approaches presented in this chapter, the basic strategy, their aim with regard to hazard and exposure of GD, major vulnerabilities as well as a rough characterization of their developmental stage are given in Table 7.2.

Summary

As the list in Table 7.2 shows, design options for HEG-GDs discussed so far are aiming at a reduction of environmental exposure. Additionally, the hazard potential of GDs will most probably be very case-specific because it is largely dependent on the genomic localization of the drive and the function of possible cargo genes. Hence, a focus on exposure minimization is justified. Potentials of exposure and hazards connected with the use of transgenes for genomic modifications can be reduced by alternative approaches based on naturally occurring phenomena with impact on the population size. All design variants of HEG-GDs with reduced risk potential are rather poorly characterized so far. Proofs of principle in a scale that enables reliable statements on their performance with regard to releases are lacking. In any case,

Table 7.2 Overview of (a) design options for HEG-GDs that may decrease their risk potential and (b) alternative approaches for population control

Technique	Main strategy	Aim (Hazard/Exposure)	Vulnerability	Remarks/ developmental stage
HEG-GD design options				
Split drive	Separation of genes for sgRNA and endonuclease	Limitation of exposure to GDO (temporal and spatial)	Co-localization of genes for Cas9 and sgRNA by recombination resulting in a global drive	Tested in yeast but no experimental proof for the limiting potential so far
Daisy chain drive/ Daisy field drive	Chain of interdependent drives/multiple separately encoded sgRNAs for endonuclease (and cargo) target sequence	Limitation of exposure to GDO (temporal and spatial)	Co-localization of genes for Cas9 and sgRNA targeting its own insertion site by recombination resulting in a global drive	No experimental proof for the limiting potential so far
(Synthetic) inductor molecule	Dependency on the supply of a substance	Limitation of exposure to GDO by GD deactivation	Germline in multicellular organisms might be difficult to target with an inductor	No exact theoretical description and no experimental proof for the deactivating potential so far
Specific (synthetic) target sequence	Targeting of a unique target sequence	Exposure limitation to GDO by targeting a genetic subpopulation	Similarity to sequences in the general population	"Synthetic site targeting" tested in yeast in laboratory scale
Environmental conditions	Self-destruction depending on environmental conditions	Limitation of exposure to GDO	Mutations deactivating the self-destruction system	No exact theoretical description and no experimental proof for the deactivating potential so far
Genetic instability	Accumulation of GD-resistant target sequences due to mutation and sequence variations	Limitation of exposure to GDO, slowdown of GD spread	Incomplete reduction of the GD frequency	First experimental observations in laboratory scale
Secondary release				
Overwriting drive	Release of secondary GD targeting the sequence of the first drive	Reducing exposure to GDO by deactivation/limitation of the initial drive and immunization of the target population	Dependence on perfect coverage of the first drive's distribution, sensitive to mutations	No experimental proof for the limiting potential so far

(continued)

Table 7.2 (continued)

Technique	Main strategy	Aim (Hazard/Exposure)	Vulnerability	Remarks/ developmental stage
gRNA targeting a drive	Release of organisms carrying gRNA against the sequence of the released GD	Reducing exposure to GDO by deactivation/ limitation of the initial drive and immunization of the target population	Dependence on perfect coverage of the first drive's distribution, sensitive to mutations	First experimental proof-of-principle in laboratory scale in *Drosophila*
Limitation by sterility	Release of sexually compatible but sterile organisms	Slowdown up to limitation of GD spread (in case of high threshold-drives)	Dependence on perfect coverage of the first drive's distribution/spread of GD is only retarded	No experimental proof for the limiting potential so far
Non-GD genetic modifications				
RIDL	Death of offspring due to dominant lethal gene	Limitation of exposure due to self-limiting approach	Mass releases of GMO required, tetracycline-dependence	Successfully applied in field trials
Resistance by transgenes	Modification of target organisms instead of suppression	Limitation of exposure due to Mendelian inheritance pattern	Evolution of resistance	First laboratory experiments showed transient effects
Sterile insect technique (SIT)	Mass release of sterile insects	Avoidance of hazard and exposure of/to GDO	Periodic mass releases required	Long experience in SIT applications (releases)
Use of naturally occurring phenomena				
Trojan female technique (TFT)	Harnessing naturally occurring male-specific mutations causing infertility	Avoidance of hazard and exposure of/to GDO/GMO	Up to now comparably small effectivity	First laboratory experiments
Wolbachia	Suppression of mosquito populations or reduction of susceptibility of mosquitoes to pathogens	Avoidance of hazard and exposure of/to GDO/GMO	Possibly increased virulence of pathogens due to selection and co-evolution	First releases already conducted

they will be difficult to achieve because with an experimental release the risk of uncontrolled spread in the case of malfunction is high.

Secondary releases of overwriting drives, gRNA targeting the sequence of a released drive or the release of sterile mating partners must be competent enough to cover all parts of a population that have been infected with the primarily released drive. It is therefore necessary to assure that mutations or a fitness burden do not interfere and reduce the intended effect of these countermeasures. A first proof-of-principle in the laboratory scale was already published in 2016 for the secondary release of a gRNA-based system as a so-called Cas9-triggered chain ablation (CATCHA) (Wu et al. 2016). However, a demonstration of the effectiveness of options for secondary releases under more realistic conditions is still pending.

Genetic modification by RIDL and in particular SIT as non-GD techniques is far better characterized, but accompanied with high efforts for the mass releases that are necessary for both. The selective application of naturally occurring phenomena with influence on population size may represent the methods of choice if transgene spread should be avoided. Moreover, approval for application will probably be much easier to obtain than for techniques based on genetic engineering. For the use of *Wolbachia* results from some first experimental releases are already available. Successful small-scale field trials have motivated larger scale releases (Burt 2014). Unfortunately, for TFT, a theoretically quite promising approach, only some first experimental results in the laboratory scale are available. From these so far rather inconclusive data, no statement can be made with regard to the potential of this advantageous technique that does not depend on transgenes or infection with parasitic bacteria. Further research is needed to characterize the suitability of using TFT as an alternative approach for population suppression.

As this comparative overview shows, alternative approaches for GDs are limited to an application for population suppression. Only an infection with *Wolbachia* bacteria can also be used to reduce the susceptibility of the host organism to pathogens and thus change the properties of a population. So far, approaches based on *Wolbachia* as "cytoplasmic drive" (Dobson et al. 2002) instead of a genetic drive represent the most advanced and functional alternative to GDs. However, regardless of the natural origin of this approach, potential impacts have to be carefully investigated. It may turn out as a highly powerful method for population suppression whose control in spread (at least for the invasive approaches) is severely limited. Moreover, a spread of *Wolbachia* to non-target species is possible. Whether TFT is able to serve as a more controllable technique in this regard is worth further investigation.

References

Akbari, O. S., Bellen, H. J., Bier, E., Bullock, S. L., Burt, A., Church, G. M., et al. (2015). Safeguarding gene drive experiments in the laboratory. *Science, 349*, 927–929. https://doi.org/10.1126/science.aac7932.

Alphey, L. (2014). Genetic control of mosquitoes. *Annual Review of Entomology, 59*, 205–224. https://doi.org/10.1146/annurev-ento-011613-162002.

Alphey, L., McKemey, A., Nimmo, D., Neira Oviedo, M., Lacroix, R., Matzen, K., et al. (2013). Genetic control of Aedes mosquitoes. *Pathogens and Global Health, 107*, 170–179. https://doi.org/10.1179/2047773213Y.0000000095.

Ant, T., Koukidou, M., Rempoulakis, P., Gong, H.-F., Economopoulos, A., Vontas, J., et al. (2012). Control of the olive fruit fly using genetics-enhanced sterile insect technique. *BMC Biology, 10*, 51. https://doi.org/10.1186/1741-7007-10-51.

Beekman, M., Dowling, D. K., & Aanen, D. K. (2014). The costs of being male: Are there sex-specific effects of uniparental mitochondrial inheritance? *Philosophical Transactions of the Royal Society B, 369*, 20130440. https://doi.org/10.1098/rstb.2013.0440.

Benedict, M., D'Abbs, P., Dobson, S. L., Gottlieb, M., Harrington, L. B., Higgs, S., et al. (2008). Guidance for contained field trials of vector mosquitoes engineered to contain a gene drive system: Recommendations of a scientific working group. *Vector Borne Zoonotic Diseases, 8*, 127–166. https://doi.org/10.1089/vbz.2007.0273.

Blagrove, M. S. C., Arias-Goeta, C., Failloux, A.-B., & Sinkins, S. P. (2012). Wolbachia strain wMel induces cytoplasmic incompatibility and blocks dengue transmission in *Aedes albopictus. Proceedings of the National Academy of Sciences of the United States of America, 109*, 255. https://doi.org/10.1073/pnas.1112021108.

Buchman, A., Marshall, J. M., Ostrovski, D., Yang, T., & Akbari, O. S. (2018). Synthetically engineered *Medea* gene drive system in the worldwide crop pest *Drosophila suzukii. Proceedings of the National Academy of Sciences*, 201713139. https://doi.org/10.1073/pnas.1713139115.

Burt, A. (2003). Site-specific selfish genes as tools for the control and genetic engineering of natural populations. *Proceedings of Biology Sciences, 270*, 921–928. https://doi.org/10.1098/rspb.2002.2319.

Burt, A. (2014). Heritable strategies for controlling insect vectors of disease. *Philosophical Transactions of the Royal Society B: Biological Sciences, 369*, 20130432. https://doi.org/10.1098/rstb.2013.0432.

Carrington, L. B., Nguyen Tran, B. C., Hoang Le, N. T., Hue Luong, T. T., Thanh Nguyen, T., Thanh Nguyen, P., et al. (2018). Field- and clinically derived estimates of Wolbachia-mediated blocking of dengue virus transmission potential in *Aedes aegypti* mosquitoes. *PNAS, 115*, 361–366. https://doi.org/10.1073/pnas.1715788115.

Champer, J., Reeves, R., Oh, S. Y., Liu, C., Liu, J., Clark, A. G., et al. (2017). Novel CRISPR/Cas9 gene drive constructs reveal insights into mechanisms of resistance allele formation and drive efficiency in genetically diverse populations. *PLOS Genetics, 13*, e1006796. https://doi.org/10.1371/journal.pgen.1006796.

Curtis, C. F. (1968). Possible use of translocations to fix desirable genes in insect pest populations. *Nature, 218*, 368–369.

Dearden, P. K., Gemmell, N. J., Mercier, O. R., Lester, P. J., Scott, M. J., Newcomb, R. D., et al. (2018). The potential for the use of gene drives for pest control in New Zealand: A perspective. *Journal of the Royal Society of New Zealand, 48*, 225–244. https://doi.org/10.1080/03036758.2017.138503.

DiCarlo, J. E., Chavez, A., Dietz, S. L., Esvelt, K. M., & Church, G. M. (2015). Safeguarding CRISPR-Cas9 gene drives in yeast. *Nature Biotechnology, 33*, 1250–1255. https://doi.org/10.1038/nbt.3412.

Dobson, S. L., Marsland, E. J., & Rattanadechakul, W. (2002). Mutualistic Wolbachia infection in *Aedes albopictus*: Accelerating cytoplasmic drive. *Genetics, 160*, 1087–1094.

Dyck, V. A., Hendrichs, J., & Robinson, A. S. (Eds.). (2005). *Sterile insect technique: Principles and practice in area-wide integrated pest management*. Netherlands: Springer.

Esvelt, K. M., & Gemmell, N. J. (2017). Conservation demands safe gene drive. *PLOS Biology, 15*, e2003850.

Esvelt, K. M., Smidler, A. L., Catteruccia, F., & Church, G. M. (2014). Concerning RNA-guided gene drives for the alteration of wild populations. *eLife, 3*. https://doi.org/10.7554/eLife.03401.

Frank, S. A., & Hurst, L. D. (1996). Mitochondria and male disease. *Nature, 383*, 224. https://doi.org/10.1038/383224a0.

Gemmell, N. J., Jalilzadeh, A., Didham, R. K., Soboleva, T., & Tompkins, D. M. (2013). The Trojan female technique: A novel, effective and humane approach for pest population control. *Proceedings of the Royal Society B, 280*, 20132549. https://doi.org/10.1098/rspb.2013.2549.

Gorman, K., Young, J., Pineda, L., Márquez, R., Sosa, N., Bernal, D., et al. (2015). Short-term suppression of *Aedes aegypti* using genetic control does not facilitate *Aedes albopictus. Pest Management Science, 72*, 618–628. https://doi.org/10.1002/ps.4151.

Hammond, A. M., Kyrou, K., Bruttini, M., North, A., Galizi, R., Karlsson, X., et al. (2017). The creation and selection of mutations resistant to a gene drive over multiple generations in the malaria mosquito. *PLoS Genetics, 13*, e1007039. https://doi.org/10.1371/journal.pgen.1007039.

Harris, A. F., Nimmo, D., McKemey, A. R., Kelly, N., Scaife, S., Donnelly, C. A., et al. (2011). Field performance of engineered male mosquitoes. *Nature Biotechnology, 29*, 1034–1039. https://doi.org/10.1038/nbt.2019.

Kyrou, K., Hammond, A. M., Galizi, R., Kranjc, N., Burt, A., Beaghton, A. K., et al. (2018). A CRISPR–Cas9 gene drive targeting doublesex causes complete population suppression in caged *Anopheles gambiae* mosquitoes. *Nature Biotechnology, 36*, 1062–1066. https://dx.doi.org/10.1038/nbt.4245..

Lacroix, R., McKemey, A. R., Raduan, N., Wee, L. K., Ming, W. H., Ney, T. G., et al. (2012). Open field release of genetically engineered sterile male *Aedes aegypti* in Malaysia. *PLOS ONE, 7*, e42771. https://doi.org/10.1371/journal.pone.0042771.

Ledford, H. (2016). Fast-spreading genetic mutations pose ecological risk. *Nature News*. https://doi.org/10.1038/nature.2016.20053.

Li, Z., Liu, Z.-B., Xing, A., Moon, B. P., Koellhoffer, J. P., Huang, L., et al. (2016). Cas9-guide RNA directed genome editing in soybean. *Plant Physiology, 169*, 960–970. https://doi.org/10.1104/pp.15.00783.

Lin, Y.-C., Wu, J.-W., & Liu, D.-P. (2012). New vector control measures on dengue fever: A literature review. *Taiwan Epidemiology Bulletin, 28*, 224–232.

Macias, V. M., Ohm, J. R., & Rasgon, J. L. (2017). Gene drive for mosquito control: Where did it come from and where are we headed? *International Journal of Environmental Research and Public Health, 14*, 1006. https://doi.org/10.3390/ijerph14091006.

McFarling, U. L. (2017). Could this zoo of mutant mosquitoes lead the way to eradicating Zika? STAT News. https://www.statnews.com/2017/12/13/gene-drive-mosquitoes-darpa/.

Min, J., Noble, C., Najjar, D., & Esvelt, K. M. (2017a). Daisy quorum drives for the genetic restoration of wild populations. *BioRxiv*. https://dx.doi.org/10.1101/115618

Min, J., Noble, C., Najjar, D., Esvelt, K. M. (2017b). Daisyfield gene drive systems harness repeated genomic elements as a generational clock to limit spread. bioRxiv preprint first posted online Feb. 6, 2017. https://dx.doi.org/10.1101/104877

Min, J., Smidler, A. L., Najar, D., & Esvelt, K. M. (2018). Harnessing gene drive. *Journal of Responsible Innovation, 5*, S40–S65. https://doi.org/10.1080/23299460.2017.1415586.

National Academies of Sciences. (2016). *Gene drives on the horizon: Advancing science, navigating uncertainty, and aligning research with public values*. Washington, DC: The National Academies Press. https://doi.org/10.17226/23405.

Neve, P. (2018). Gene drive systems: Do they have a place in agricultural weed management? *Pest Management Science, 74*, 2671–2679. https://doi.org/10.1002/ps.5137.

Noble, C., Min, J., Olejarz, J., Buchthal, J., Chavez, A., Smidler, A. L., et al. (2019). Daisy-chain gene drives for the alteration of local populations. *Proceedings of the National Academy of Sciences* 1–8. https://doi.org/10.1073/pnas.1716358116.

Oye, K. A., Esvelt, K., Appleton, E., Catteruccia, F., Church, G., Kuiken, T., et al. (2014). Regulating gene drives. *Science, 345*, 626–628. https://doi.org/10.1126/science.1254287.

Phuc, H. K., Andreasen, M. H., Burton, R. S., Vass, C., Epton, M. J., Pape, G., et al. (2007). Late-acting dominant lethal genetic systems and mosquito control. *BMC Biology, 5*, 1–11. https://doi.org/10.1186/1741-7007-5-11.

Regalado, A. (2017). *Farmers seek to deploy powerful gene drive*. MIT Technology Review Accessed 12.13.17.

Rodriguez-Beltran, C. (2012). GM mosquitoes: Survival in the presence of tetracycline contamination. TWN Biosafety Briefing, 16.02.12, https://www.biosafety-info.net/article.php?aid=878. Accessed 09.08.17.

Schliekelman, P., & Gould, F. (2000). Pest control by the release of insects carrying a female-killing allele on multiple loci. *Journal of Economic Entomology, 93*, 1566–1579.

Simon, S., Otto, M., Engelhard, M. (2018). Synthetic gene drive: between continuity and novelty. *EMBO Rep. e45760*. https://doi.org/10.15252/embr.201845760.

Subbaraman, N. (2011). Science snipes at oxitec transgenic-mosquito trial. *Nature Biotechnology, 29*, 9–11.

Thomas, D. D., Donnelly, C. A., Wood, R. J., & Alphey, L. S. (2000). Insect population control using a dominant, repressible. *Lethal Genetic System Science, 287*, 2474. https://doi.org/10.1126/science.287.5462.2474.

Weinert, L. B., Araujo-Jnr, E. V., Ahmed, M. Z., & Welch, J. J. (2015). The incidence of bacterial endosymbionts in terrestrial arthropods. *Proceedings of Biology Science, 282*, 20150249. https://doi.org/10.1098/rspb.2015.0249.

Wolff, J. N., Gemmell, N. J., Tompkins, D. M., & Dowling, D. K. (2017). Introduction of a male-harming mitochondrial haplotype via 'Trojan Females' achieves population suppression in fruit flies. *eLife, 6*, e23551. https://doi.org/10.7554/eLife.23551.

Wu, B., Luo, L., & Gao, X. J. (2016). Cas9-triggered chain ablation of cas9 as a gene drive brake. *Nature Biotechnology, 34*, 137–138. https://doi.org/10.1038/nbt.3444.

Wright, O., Stan, G.-B., and Ellis, T. (2013). Building-in biosafety for synthetic biology. *Microbiology, 159*, Pt 7, 1221–1235. https://doi.org/10.1099/mic.0.066308-0.

Zentner, G. E., & Wade, M. J. (2017). The promise and peril of CRISPR gene drives. *BioEssays, 39*, 1700109. https://doi.org/10.1002/bies.201700109.

8

Gene Drive Organisms and Risk Analysis: Challenges and Complexities

Christoph Then

Introduction

People have selected and cross-bred plants (and animals) for thousands of years in order to establish beneficial and desirable traits. However, natural mechanisms such as gene regulation and heredity can now be circumvented with modern technical tools of genetic engineering. In consequence, experience gained from conventional plant breeding cannot simply be extrapolated to the risk assessment of GE plants. According to EU regulation (Directive 2001/18), all organisms derived from processes of genetic engineering require a risk assessment before they can be released.

New challenges have arisen with applications such as 'gene drives' that are intended to be introduced into natural populations and give rise to offspring that spread and propagate throughout those populations. Gene-drive mechanisms were, for example, successfully established in laboratory populations of *Drosophila* by using the nuclease CRISPR/Cas9 (Gantz and Bier 2015). Such organisms replicate the process of genetic engineering in a self-organised way: in every generation the nuclease is meant to copy and insert itself at a given location within the genome. This process is also named 'mutagenic chain reaction' (see Gantz and Bier 2015; Ledford 2015). As a result, the newly introduced DNA can spread through a population exponentially, and much more rapidly than could be expected under the Mendelian pattern of inheritance.

The risk assessment of a potential release of a gene drive organism into the environment needs to consider uncertainties and limits of knowledge on at least three levels: the technology, the target organism and the receiving environment, including abundant non-target organisms (NTOs). Moreover, methodological problems need to be overcome: the comparative approach that is the starting point for current EFSA (EFSA 2010) environmental risk assessment might not be applicable due to the lack

C. Then (✉)
Testbiotech e.v, Munich, Germany
e-mail: christoph.then@testbiotech.org

of suitable 'comparators'. The following paragraphs therefore discuss criteria and methodologies that can be applied in the risk analysis of gene drive organisms in the face of substantial uncertainties.

The Production of Knowledge and Non-knowledge

There can be various reasons for non-knowledge. It can, for example, be due to unawareness of facts that are already known in a specific field of expertise. This kind of non-knowledge can be easily remedied. Other areas of non-knowledge and the production of non-knowledge can create large and systemic problems. In some circumstances, we might not even be aware of our limits of knowledge.

The Science of Non-knowledge in Upstream Technology Assessment

The science of non-knowledge already has a long tradition in the context of technology assessment. Some of the early debates were triggered by discussions around nuclear power and chemical pollution. This report discusses some selected aspects.

In 1992, Wynne developed a classification system based on the criteria "risk", "uncertainty", "ignorance" and "indeterminacy" as summarised in Table 8.1.

These criteria are explained by Wynne (1992) in more detail (for all quotes in bullet points see page 114):

- The term risk can be applied when "the system behaviour is basically well known, and chances of different outcomes can be defined and quantified by structured analysis of mechanisms and probabilities."
- The term uncertainty can be applied "if we know the important system parameters but not the probability distributions (…). These uncertainties are recognized, and explicitly included in analysis."
- According to Wynne "a far more difficult problem is ignorance, which by definition escapes recognition. This is not so much a characteristic of knowledge itself as of the linkages between knowledge and commitments based on it—in effect, bets

Table 8.1 Classifications established by Wynne (1992)

Risk	Know the odds
Uncertainty	Don't know the odds: may know the main parameters, may reduce uncertainty but increase ignorance
Ignorance	Don't know what we don't know, ignorance increases with increased commitment based on current knowledge
Indeterminacy	Causal chains or networks open

(technological, social, economic) on the completeness and validity of that knowledge. (…) The conventional view is that scientific knowledge and method enthusiastically embrace uncertainties and exhaustively pursue them. This is seriously misleading. It is more accurate to say that scientific knowledge gives prominence to a restricted agenda of defined uncertainties—ones that are tractable—leaving invisible a range of other uncertainties, especially about the boundary conditions of applicability of the existing framework of knowledge to new situations. Thus ignorance is endemic to scientific knowledge, which has to reduce the framework of the known to that which is amenable to its own parochial methods and models."
- Finally, "indeterminacy exists in the open-ended question of whether knowledge is adapted to fit the mismatched realities of application situations, or whether those (technical and social) situations are reshaped to 'validate' the knowledge."

The Exploration of Non-knowledge in the Field of Biotechnology

Böschen (2006, 2009) and his colleagues explored areas of non-knowledge, especially in the context of biotechnology. Böschen et al. (2006) put forward three dimensions of non-knowledge (p. 297):

- "The first dimension refers to knowledge (or awareness) of non-knowledge, which spreads between full awareness of non-knowledge (we know what we don't know) and complete unawareness ('unknown unknowns').
- The second dimension, intentionality of non-knowledge, contrasts unintended non-knowledge with the conscious refusal of certain cognitions.
- The third dimension, temporal stability (or reducibility) of non-knowledge, extends from what is not yet known, but (presumably) does not present any substantial difficulties to cognition, to the entirely 'unknowable' and therefore uncontrollable."

Based on these criteria, the perspective of scientists in the field becomes important and allows exploration of what is known as 'cultures of non-knowledge'. Böschen et al. (2006) show that two interrelated scientific disciplines—molecular biology and ecology—"entail different types of non-knowledge and deal with non-knowledge differently (…) The scientific culture of non-knowledge in molecular biology can be described as control-oriented, while that of ecology can be described as uncertainty-oriented." (p. 295).

In regard to the boundaries of knowledge, Böschen (2009) argues that one decisive question remains to be answered in order to implement the PP as requested by EU regulations e.g. 178/2002 and Directive 2001/18 (p. 509): "Although there is an institutional solution, one all-decisive question is remaining. What is the actual evidence on which decisions about the applicability of the PP are to be taken? And which evidence is necessary to decide about different precautionary strategies? These questions are difficult to answer with respect to the debate about non-knowledge. But they have to be answered, because political decisions are always based on knowledge (…).”

Adequate management of non-knowledge in implementing the PP cannot be established if Böschen’s (2009) question cannot be answered. Consequently, within the regulatory processes, inherent non-knowledge might increase unnoticed to such an extent that sufficiently robust risk assessment is disabled.

Böschen (2009) shows there are basically two options in regard to performing risk assessment in this context: a “restrictive evidential culture (e.g., molecular biology)” and a “holistic evidential culture (e.g., ecology)”. These are accomplished by a third approach which could be seen as a pragmatic compromise: “evaluative evidential cultures (e.g., environmental medicine)”.

Böschen (2009) summarises (p. 513): “Control-oriented epistemic cultures proceed in a restrictive-experimental way and are oriented towards an improvement of (technological) options for action. In contrast, complexity-oriented epistemic cultures structure their knowledge in a holistic-contextual way and enhance options for reflection. Finally, expertise-based epistemic cultures are marked by the combination of diagnostic knowledge and knowledge on problem-solving. There, epistemic strategies are related towards an improvement of options for decisions. All of these cultures generate knowledge relevant for making decisions, but they do not find a balanced attention in the risk policy of the GMOs. Therefore, a selection process of the knowledge resources relevant for the conflicts occurs.”

Building on arguments and criteria developed by Böschen and his colleagues (2006/2009), one can conclude that it is necessary to integrate adequate management of non-knowledge into the field of genetic engineering and biotechnology in upstream prospective technology assessment, as well as within ‘end of pipe’ regulatory decision-making for specific products and organisms.

Some ‘Known Unknowns’ in Regard to Risk Assessment of GE Organisms and New Challenges Posed by Gene Drives

Decisions made by risk managers are always dependent on plausibility and knowledge and cannot be based on speculation. Risk assessment has to be organised in a way that the final decision-making of the risk manager is sufficiently informed as to whether a GE organism can be considered to be safe and therefore allowed for release. However, due to the complexity of the associated mechanisms and various interactions between the target organisms and their environment, substantial uncertainties and areas of non-knowledge have to be taken into account when it comes to the risk assessment of gene drive organisms.

In this context, two questions are of crucial relevance:

What are the ‘known unknowns’ stemming from experience with already existing GE organisms that give ‘reasons for concern’?

What are the main challenges in risk assessment of gene drives that go beyond the experience with existing GE organisms?

To explore these questions, some conceptual challenges of GE organism risk assessment are given as a starting point. Subsequent sections set out a more detailed investigation of the differences between GE plants and plants derived from traditional breeding. This is followed by a discussion of questions arising from the risk assessment of gene drives.

Conceptual Challenges in Risk Assessment of GE Organisms

As a starting point for the discussion on the risk assessment of GE organisms, it is useful to consider what is often called the complexity of biology. A comparison to the risk assessment of chemicals might be quite useful in this context: while chemicals (in many cases) can be considered clearly defined entities, the characteristics of organisms are largely shaped by interactions and the mechanisms of self-reproduction, self-organisation and adaptability. Conceptual challenges for the risk assessment of GE organisms can be identified on several levels.

What is the Entity that Has to be Assessed?

To some extent, life forms can only be assessed in combination with their environment: for example, the well-established concept of the 'holobiont' (see for example Richardson 2017) shows that multicellular organisms such as plants, insects or mammals can hardly be separated from their associated microbiome. The organism and its associated microbiome interact very closely: it is known that the microbiome can extensively impact the biological characteristics and health status of humans, plants and animals (see for example Lynch and Pedersen 2016; da Silva et al. 2016).

How to Assess Complex Cause-Effect Relationships?

Well defined cause–effect relationships may frequently not be applicable in the case of life forms: as can be shown in GE plants (for some references see below in 3.3), the interaction of the inserted genes with the genetic background as well as the interactions of the organisms with their environment can play an important role. These interactions can create effects in a bi-directional and non-linear manner: it is not only the organisms that impact the environment, the various environmental conditions, abiotic and biotic stressors also impact the biological characteristics of organisms. Thus, risk assessment of GE organisms not only has to assess the impact of the organisms on the environment, but also vice versa. In addition, the resulting combinatorial effects also have to be taken into account.

Will the Characteristics of the Relevant Entities Remain Predictable in Future?

The characteristics of GE organisms might change from one generation to the next (for some references see below in 3.3). With self-organisation and self-reproduction and in interaction with changing environmental conditions, next generation effects may occur that cannot be predicted on the level of the previous generations. Even if DNA is transmitted to the next generation in a way that genetic stability is assumed on the genomic level, this does not mean that the intended function of the gene and the associated phenotype will be transmitted to the offspring as well. Thus, next generation effects have to be considered in all cases where GE organisms might be able to persist and propagate in the environment. This is especially relevant if gene flow occurs from the GE organism into wild populations.

How to Take Communication and Signalling Pathways Between Organisms into Account?

Life forms interact with the environment via multiple bio-chemical pathways. In plants, these pathways include signalling and communication with other plants, microorganisms and beneficial insects (see for example Schaefer and Ruxton 2011). There are various compounds involved such as volatile substances, other secondary metabolites and biologically active compounds. Environmental risk assessment of GE organisms should include the various ways in which organisms interact and communicate with their environment, and these might not be well defined in all cases.

The conceptual challenges listed above show that, given the terminology introduced by Böschen (2009), a "restrictive evidential" approach" is not sufficient to perform risk assessment on GE organisms, and a more "holistic evidential" approach" has to be applied.

Specific Characteristics of GE Organisms

Risks of GE organisms differ from those associated with organisms derived from natural evolutionary processes. In the overview we have used plants to show some relevant differences; this is because most of the experience we have has been obtained from the environmental risk assessment of GE organisms in regard to plants.

Conventional plant breeding can look back on long-standing experience. The mechanisms used are generally based on the methods and results of evolutionary processes, such as selection: plant breeding starts from a broad range of biodiversity that is used for selection and is newly combined through crossing. In addition, since ca. the 1950s, it has been possible to enhance genetic variation by technically inducing mutations. These methods are known as mutagenesis (Oladosu et al. 2016) and do not profoundly change the pattern of emerging genetic variations; they more or less simply speed up evolutionary processes that might also occur naturally.

In short, the methods and mechanisms used in what is known as 'conventional' breeding:

- make use of huge genetic diversity as a starting point;
- are applied to the whole cell or organisms;
- do not insert or delete genetic information targeted by technical means.

Therefore, conventional breeding does not change the mechanisms of natural heredity and gene regulation. Furthermore, in many cases, each step might be relatively small in regard to the relevant plant characteristics. Thus, plant characteristics are improved by breeding processes that can take many years and involve many varieties. This is a constantly ongoing process that allows breeders to gain experience with each of the specific traits over a longer period of time. Nevertheless, some organisms resulting from conventional breeding might require risk assessment in regard to health and the environment. For example, it is possible to establish herbicide resistant crop plants by means of conventional breeding, which should be investigated in regard to their impact on weedy species and biodiversity (Burgos et al. 2014).

On the other hand, genetic engineering is based on "techniques involving the direct introduction into an organism of heritable material prepared outside the organism" (Directive 2001/18, Annex I A). These techniques and processes show technical characteristics that are distinct from those of conventional breeding:

- The techniques applied in genetic engineering allow mechanisms of natural heredity and gene regulation to be by-passed.
- Direct intervention on the level of the genome means that traits can be established that do not occur naturally e.g. plants which produce insecticidal proteins derived from *Bacillus thuringiensis* (Bt).
- In many cases, the additionally inserted genes are not identical to those found in nature: for example, in the case of plants that produce Bt toxins, the DNA sequences are modified in the laboratory giving rise to truncated or chimeric Bt proteins that do not exist in nature (see Hilbeck and Otto 2015).

The biological changes in regard to plant characteristics can in many cases be extensive and might even be considered 'disruptive'. The resulting plants might be cultivated on large scale without any experience being gathered over a longer period of time.

In summary, experience gained from conventional plant breeding cannot simply be extrapolated to the risk assessment of GE plants. Thus, according to EU law (Directive 2001/18), all organisms derived from processes of genetic engineering generally require risk assessment before they are released into the environment.

Environmental risk assessment (ERA) of GE plants examines several aspects (EFSA 2010): it encompasses the trait (such as the Bt toxin), the organism and its genetic stability (including gene expression, stability of the gene functions) and the interactions of the organism with the receiving environment on various levels.

Specific Challenges in Risk Assessment of Gene Drives

There are several genetic engineering techniques that can be used to create a gene drive, many of them are based on applications of the nuclease CRISPR/Cas. There is some proof of principle that gene drive can be established in yeast (Di Carlo et al. 2015), mosquitoes (Gantz and Bier 2015; Hammond et al. 2015; Kyrou et al. 2018), flies (Champer et al. 2017; KaramiNejadRanjbar et al. 2018; Buchman et al. 2018) and mice (Grunwald et al. 2018). Each of these outcomes will require a case by case risk assessment in regard to its technical characteristics, the target species and the receiving environment (Akbari et al. 2014; Kuzma et al. 2017; Noble et al. 2017; Oye et al. 2014).

In addition, there are some general characteristics that can be used to distinguish gene drive organisms from other GE organisms that have been assessed by EU institutions so far:

- Where gene drives are based on CRISPR/Cas, the process of genetic engineering becomes inherited and self-replicating in subsequent generations; established as a self-organising process, largely outside of efficient or ongoing control mechanisms.
- Changes in the 'laws' of inheritance are so fundamental that in many cases it will hardly be possible to find suitable comparators, even though this is a requirement of the 'comparative risk assessment' approach.
- Gene drives have been developed specifically to genetically engineer species that are non-domesticated. Consequently, the additional genetic information will be introduced into a wider range of genetic backgrounds which—especially in wild, natural populations—can be quite heterogeneous and give rise to a wide range of unexpected effects (see also Chandler et al. 2013; Mullis et al. 2018; Evangelou et al. 2018; Saltz et al. 2018).
- Gene drives are intended to target wild, natural populations, therefore, a wider range of possible interactions with the receiving environment and the ecosystems has to be expected.
- If gene drives are introduced into wild, natural populations, it can become much more difficult to intervene if adverse effects emerge than with crops grown in the fields.

Some of the differences in risk assessment between gene drive organisms and GE crop plants are summarised in Table 8.2.

In conclusion, the hypothesis discussed in the following chapters is that gene drives and also other GE organisms that can persist and propagate in the environment and/or enable gene flow to wild populations, pose new challenges for EU risk assessment. What can be expected is a substantial increase in spatio-temporal complexity and a decrease in the robustness of overall risk analysis (see also Simon et al. 2018).

Table 8.2 Some new challenges in the risk assessment of GE gene drive organisms in comparison to GE crop plants

Some aspects of the risk assessment of GE crop plants	New challenges in the risk assessment of gene drive organisms
The majority of crop plants are cultivated for only one growing period. These plants are not meant to reproduce spontaneously	Next generations will emerge spontaneously; the process of genetic engineering is a self-organised process replicating in each generation
Due to previous breeding processes, plant varieties as used for genetic engineering, are stable and have defined characteristics, as well as reduced genetic diversity. Seed quality can be controlled by breeders (or farmers) before and during cultivation	Wild populations very often inherit a broad spectrum of heterogeneous genetic backgrounds. As a result, gene drive organisms can introduce their new genetic information into heterogeneous genetic backgrounds without additional controls in place
Crop plants of the same species are often cultivated under similar environmental conditions in a managed agricultural system	Wild populations e.g. insects are often exposed to a wider range of environmental conditions due to their mobility. Further impact factors include e.g. seasonal changes
Crop plants are often grown in an environment of agricultural systems with reduced biodiversity	Wild populations very often interact with complex ecosystems

Some Reasons for Concern Arising from Existing Evidence

To test the hypothesis, it has to be investigated whether there is any supporting evidence that.

- the process of spontaneous self-reproduction of GE organisms increases uncertainty regarding genetic stability (including gene expression and stability of the gene functions) in the offspring generations;
- a higher range of genetic diversity within the target populations increases uncertainty regarding genetic stability (including gene expression and stability of the gene functions) in the offspring generations;
- interaction with a more complex environment increases the likelihood of unexpected effects in GE organisms.

To answer these questions, research was conducted within peer reviewed publications on the risk assessment of existing GE crop plants in the EU. Some of this research was aimed at GE plants that had unintentionally escaped into natural populations or had started to become feral (see, for example, Bauer-Panskus et al. 2013). Other relevant publications were those dealing with the responses of GE plants to changes in environmental conditions (see, for example, Zeller et. al. 2010). Based on the existing publications, evidence can be established for relevant aspects. Some of these findings are summarised in Table 8.3. Some examples are explored in more detail below. It was concluded that existing experience was in line with the above hypothesis.

Table 8.3 Existing experience with GE organisms with specific relevance for risk assessment of gene drives organisms

Topic	Findings
Next generation effects	Next generations of GE organisms can show effects that were not observed or intended in the original event (Kawata et al. 2009; Cao et al. 2009; Yang et al. 2017)
Effects emerging from genetic background	Unintended effects can emerge from interaction of the newly inserted genes with the genetic backgrounds (Bollinedi et al. 2017; Lu and Yang 2009; Vacher et al. 2004; Adamczyk and Meredith 2004; Adamczyk et al. 2009)
Interaction with the environment on the level of the genome (genome × environment interactions)	Unintended genomic effects can be triggered by changing environmental conditions or biotic and abiotic stressors (Zeller et al. 2010; Matthews et al. 2005; Meyer et al. 1992; Trtikova et al. 2015; Then and Lorch 2008; Zhu et al. 2018; Fang et al. 2018)

These findings can be explained in more detail by exploring some of the examples, whereby the process and result of genetic engineering can be assumed to be a cause for the unintended effects:

- Bollinedi et al. (2017) crossed lines of so-called "Golden Rice" with the Indian variety Swarna and observed growth disturbance since the gene constructs interfered with the plant's own gene for producing growth hormone. Further, the gene constructs were not, as intended, active solely in the kernels, but also in the leaves. This led to a substantial reduction in the content of chlorophyll that is essential for vital functions in the plants. This effect was not observed in other varieties. Genetic background interaction is a commonly observed phenomenon in many species (see, for example, Table 8.3 for further examples).
- Fang et al. (2018) showed that higher fitness does occur in GE glyphosate resistant plants in a glyphosate-free environment. According to this research, the enzyme EPSPS (5-enolpyruvylshikimate-3-phosphate synthase) produced in the plants not only makes the plants resistant to glyphosate, it also interferes with plant metabolism for growth and fecundity. As a consequence, the offspring of the plant can produce more seeds and be more resistant to environmental stressors such as drought and heat. They also describe the interaction between the genome and the environment: for Arabidopsis producing additional EPSPS enzymes it was observed that seed germination ratios increased significantly when transgenic seeds were exposed to heat and drought stressors, although no differences were found in seed germination among different lines when seeds were exposed to normal temperatures.
- Transgenic oilseed rape is known to have become established independently of cultivation in several regions of the world, such as Canada, the US, Japan, Australia

and Switzerland (Bauer-Panskus et al. 2013). Interestingly, some populations seem to be self-sustaining and can persist without additional gene flow (spillage) at Japanese harbors (Katsuta et al. 2015). Similar findings were also reported from Canada (Warwick et al. 2008; Knispel and McLachlan 2010). This is a strong indication that fitness of the offspring of the transgenic plants was underestimated. Next generations effects were also observed in rice, especially if crossed with weedy rice: Cao et al. (2009) describe crosses between insecticidal rice with weedy rice that causes taller plants, more tillers, panicles and spikelets per plant, as well as higher seed weight, compared with the weedy rice parents. Seeds from the F1 hybrids had higher germination rates and produced more seedlings than the weedy parents.

It was concluded that the risk assessment or technical characterisation of GE organisms (plants) established in the laboratory or under controlled conditions cannot be seen as sufficient to predict all the relevant effects that can emerge in the next generations, and in interaction with the receiving environments. Therefore, parallel to an increase in spatio-temporal complexity, a decrease in the robustness of overall risk analysis is very likely.

The EFSA Concept and the Problem of Spatio-Temporal Complexity

The EFSA is the EU regulatory authority responsible for assessing the risks of GE organisms in regard to health and the environment. There is existing experience to show the way in which EFSA deals with the reasons for concern presented here. Further, the question arises of whether relevant issues can escape the current system.

The Current EFSA System and Its Approach to Future Applications

The Commission Implementing Regulation 503/2013 is applied in EFSA risk assessment of GE plants for import and usage as food and feed. It foresees a number of investigations to be performed and several sets of data to be presented by the companies. The way in which EFSA puts the regulation into practice can be regarded as more "restrictive evidential" than "holistic evidential" (using the terminology of Böschen 2009): in most cases, EFSA does not discuss limits of knowledge, and where there are uncertainties, EFSA does not generally ask for further data. For example, very often several significant findings regarding changes in plant composition are identified (see for example, EFSA 2018). However, EFSA does not see the need to request further investigations as long as there is no evidence that the changes in composition can cause harm to health or the environment (Testbiotech 2018a). Thus, risk assessment currently performed by the EFSA is largely based on a restrictive paradigm of evidence, or more broadly speaking, works within categories and systems in which risks can be quantified by structured analysis of mechanisms and probabilities, without giving sufficient weight to uncertainties and limits of knowledge.

A key element in the risk assessment of GE organisms is the comparative approach which is integrated in Annex II of Directive 2001/18/EC (see Section 2.2). As applied by EFSA (EFSA 2010), it requires the identification of differences between the GE organisms and their adequate comparator(s), of both intended and unintended effects.

Clearly there are some problems with the comparative approach. Some relevant information can be gained from comparison with wild species. However, organisms inheriting a gene drive able to produce offspring that can spread and propagate further by overriding the pattern of Mendelian inheritance, can cause changes in population dynamics and interaction with the environment that go far beyond what can be observed in the wild species.

In this context, it is notable that EFSA developed specific guidance for the environmental risk assessment of GE animals (EFSA 2013) that also addresses the issue of gene drive. So far, this guidance has not been used because there have been no applications for GE animals. It can be concluded from the published guidance (EFSA 2013) that EFSA assumes the risk assessment of GE animals and insects does indeed lead to a higher level of uncertainties in comparison to the risk assessment currently established for GE crop plants (EFSA 2010). For example, EFSA requests the applicant to consider several degrees of uncertainty:

> The formal analysis should address three broad types of uncertainty:
>
> 1. Linguistic uncertainty (…)
> 2. Variability—caused by fluctuations or differences in a quantity or process, occurring over time, with location or within a group. (…)
> 3. Incertitude—due to limitations of scientific knowledge and knowledge production systems (…). (p. 42)

Further, in comparison to the assessment of crop plants, EFSA (2013) raises additional questions. In regard to target organisms, issues such as genetic background and the life cycle are mentioned. EFSA also addresses spatio-temporal complexity:

> applicants should consider and discuss breeding in which the recombinant DNA could be introduced or introgressed into genetic backgrounds of domesticated, bred and wild individuals. (p. 25).
>
> applicants should consider the whole life cycle of the GM animal and the receiving environments of the different life stages to determine possible adverse effects over time. (p.39).
>
> long-term effects may also occur due to increases in spatial and temporal complexity. (p.39).

Furthermore, relevant issues in regard to NTOs (and the environment) include the ecological functions of specific species and their complex biotic or abiotic interactions:

> i. The ecological functions of specific species and their complex biotic or abiotic interactions (…) are not always fully understood.

ii. The methodologies for testing potential effects on NTOs are limited. Field trials might not be feasible in all cases, as it might be impossible to eradicate the released GM insect population if an adverse effect is identified related to the release, in particular, applying replacement strategies.

iii. The fact that it is not feasible to simulate the complexity of the receiving environments in laboratory tests, semi-field tests or modelling. (…) Consequences of the decrease or eradication in population size of a certain species or the replacement of wild population by GM insect populations might not be predictable. (p. 103).

EFSA also addresses the problem finding an adequate "choice of comparators" (p. 25). In summary, EFSA (2013) concurs with the hypothesis that the risk assessment of GE organisms carrying a gene drive poses new challenges in comparison to current EU risk assessment.

Some Relevant Aspects of Spatio-Temporal Complexity

It can be concluded that the following questions must be answered in respect to the risk assessment of GE organisms inheriting a gene drive and other GE organisms that can persist and propagate in the environment, including in regard to spatio-temporal complexity (see Table 8.4):

1. Can genetic stability be controlled in following generations?
2. How can genetic diversity in the target population be taken into account?
3. Will there be any gene flow to other species?
4. How can the population dynamics and life cycle aspects of the target species be integrated?
5. Can the receiving environment be defined in regard to relevant interactions and confined in regard to potential spread?

Table 8.4 indicates that in many cases significant uncertainties remain and some unknowns might prevail that make the risk assessment inconclusive: the multiplex interrelations with the closer and wider environment pose a real challenge for the risk assessor. An even bigger problem is caused by the necessity of thoroughly assessing all of the offspring generation: while genetic stability over several generations might be demonstrated in the laboratory, genome × environmental interactions and introgression into heterogeneous genetic backgrounds can still trigger unpredictable next generation effects. Whatever the case, the technical characterisation of gene drive organisms or experiments carried out in the laboratory cannot be regarded as sufficient to predict all relevant effects that can emerge in the next generations, and in interaction with the receiving environments.

This could prompt a strategy to address the remaining uncertainty within environmental monitoring. However, if monitoring reveals undesirable effects, it might not be the possible to remove the organism from the environment as it was assumed to be compatible with crop plants. In this case, EU Regulation 2001/18 foresees the possible withdrawal of authorisation. This is not likely to be effective for species which

Table 8.4 Overview of some questions relevant for risk assessment of GE gene drive organisms in regard to spatio-temporal complexity

Question	Relevance	Which methodology is available?
(1) Can genetic stability be controlled in following generations?	Self-replication and environmental as well as epigenetic effects can lead to emergence of next generation effects not observed in the first generation	Several generations should be observed under a wide range of defined environmental conditions. The outcome has to be put in context to questions (2) and (3)
(2) How can genetic diversity in the target population be taken into account?	In most cases a high degree of genetic diversity exists in natural populations. These heterogeneous genetic backgrounds can trigger unexpected effects not observed in the lab populations	In most cases the inserted genes cannot be tested in interaction with the real genetic diversity within natural populations. For example, in insects, the strains reared in the lab might only represent a small selection of the real genetic diversity within wild populations
(3) Will there be any gene flow to other species?	If gene flow is possible and hybrid offspring are viable, the resulting organisms have to be seen as new events that have to be assessed separately from the original GE organisms	It might be possible to perform crosses under controlled conditions. Results have to put in context with question (1) and (2)
(4) How can population dynamics and life cycle aspects of the target species be integrated?	Bottlenecks in the population dynamics, for example, due to the winter season, might result in inbreeding and changes in genetic variability. Bottlenecks can have a significant impact on tipping points within the populations	Large scale population effects can be modelled, but empirical investigations are difficult. Further, any results have to be interpreted in the light of question (1) and (2)
(5) Can the receiving environment be defined in regard to relevant interactions and confined in regard to potential spread?	Adverse effects can emerge from interaction with closer (associated microbiomes) or wider environments (such as food webs, predators, beneficial organisms). Terrestrial and aquatic systems have to be taken into account, as well as complex interrelations (such as signalling pathways) and behavioural aspects. Further, interrelations may vary greatly throughout the life cycle (different developmental stages such as egg, larva, pupa, adult)	These aspects have to be assessed case by case and step by step. In most cases, long-term, cumulative and combinatorial effects cannot be tested or investigated ex ante

are self-reproducing and self-sustaining. Therefore, monitoring might be useful but cannot be regarded as sufficient to address the problems outlined above.

Problems Emerging from Spatio-Temporal Complexity for Risk Assessment

If the spatio-temporal dimension cannot be defined, risk assessment has to consider evolutionary dimensions. The problem: evolutionary dynamics combine large numbers of individuals on the population level and singularities on the molecular scale. Thus, evolutionary processes make it possible to turn events with a low probability of ever happening into events that may feasibly happen (Breckling 2013). Under these conditions, for example, the fitness of new genomic constituents cannot be calculated in absolute terms; it will depend on the environment and future changes. Such evolutionary processes can cause major problems in regard to GE organisms regulated under EU Directive 2001/18. It has to be concluded that a sufficient and robust risk assessment of GE organisms can only be conducted if it is based on a spatio-temporal dimension that is clearly confined.

It is evident that, in the context of gene drives organisms, the spatio-temporal dimension is a much more pressing concern in comparison to GE plants only grown for one season. Consequently, the environmental risk assessment of GE organisms that can persist and propagate in the environment and especially of 'gene drive organisms' will result in an increasing level of uncertainty, depending on the relevant spatio-temporal dimension. At some stage, the level of uncertainties might increase to an extent that the delicate balance between knowledge and non-knowledge is distorted allowing tipping points to be reached in risk assessment, if inherent non-knowledge increases to an extent that robust risk assessment is disabled. This problem is illustrated in Fig. 8.1.

'Spatio-Temporal Controllability' as a Cut-Off Criterion

Therefore, coming back to the question posed by Böschen (2009) about "actual evidence on which decisions about the applicability of the PP are to be taken": Is it possible to categorise non-knowledge and uncertainties in a way that decision-making can be based on sufficiently clear criteria? How can non-knowledge, uncertainties, or as EFSA (2013) puts it, "incertitude, caused by limitations of scientific knowledge and knowledge production systems" be integrated into a regulatory system of decision-making? In other words, how can we create sufficient knowledge to facilitate decision-making when faced with substantial non-knowledge?

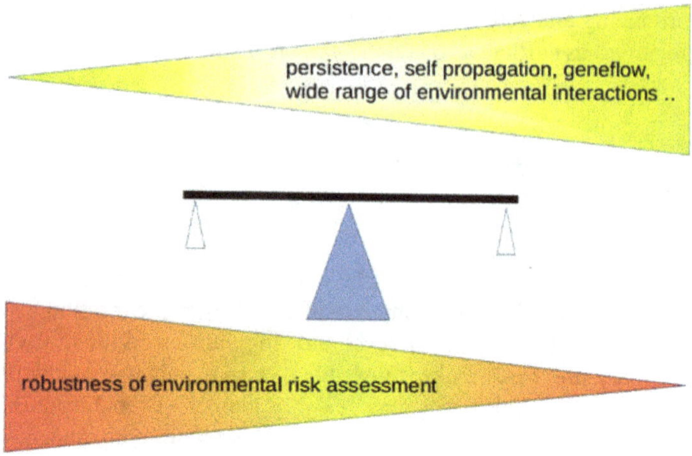

Fig. 8.1 The balance between spatio-temporal complexity and the reliability of risk assessment

Lessons Learned from Risk Assessment of Chemicals

The spatio-temporal dimension also plays a role in the risk assessment of chemical substances. For example, Recital 76 of EU Regulation 1907/2006 (REACH) addresses the issue: "Experience at international level shows that substances with characteristics rendering them persistent, likely to bioaccumulate and toxic, or very persistent and very likely to bioaccumulate, present a very high concern, while criteria have been developed allowing the identification of such substances." Consequently, criteria to identify persistent, bio-accumulative and toxic, as well as very persistent and very bio-accumulative chemical substances, are defined in ANNEX XIII of the regulation EU Regulation 1907/2006.

Further, EU Regulation 1107/2009 which concerns pesticides, integrates the criteria of POP (persistent organic pollutant), PBT (persistent, bio-accumulative, toxic) and vPvB (very persistent, very bio-accumulative) into the regulatory decision making process. These criteria function as so-called cut-off criteria: in essence, the approval process should not proceed if the substance is "POP", "PBT" or "vPvB". In this context, it is important that the chemical substances are not only assessed in regard to their toxicity but also, more generally, in regard to their "fate and behaviour in the environment" (EU Regulation 1107/2009, Annex II, 3.7.), which gives decisive weight to the spatio-temporal dimension: if a substance is regarded as very persistent and very bio-accumulative, there might still be some uncertainty or non-knowledge in regard to its actual long-term adverse effects. Nevertheless, according to EU Regulation 1107/2009, it cannot be approved.

Cut-Off Criteria in the Risk Assessment of GE Organisms

The way in which cut-off criteria were established for chemicals (including pesticides) could also be useful as a model for the risk assessment of GE organisms, especially gene drives. Similar to EU regulation of chemicals, the fate and behaviour of the organisms in the environment would be a crucial aspect. Therefore, if it were known that GE organisms could escape 'spatio-temporal controllability' by reproducing within natural populations without any effective control of spread or persistence, then the authorisation process could not proceed and the release of the GE organisms could not be allowed.

Thus, in effect, GE organisms could neither be approved nor released, including when actual long-term effects could not be determined in detail. How then can criteria be developed for the risk assessment of GE organisms that are sufficiently well defined and applicable in the approval process, as well as take into account uncertainties and limits of current knowledge? As described above in the context of chemical substances, the cut-off criteria are defined so that known characteristics of the substances are used to integrate uncertainties around actual long-term impacts into decision making.

In close analogy, the criteria applied in the risk assessment of GE organisms should be as clear and well defined as possible. Well-established scientific criteria from three areas of knowledge should be taken into consideration: (1) the (natural) biology of the target organisms (2) their (naturally) occurring interactions with the environment (biotic and abiotic) and (3) the intended technical characteristics (traits) inserted through genetic engineering. These criteria should be combined to establish an extra step in the risk assessment of GE organisms aimed at assessing 'spatio-temporal controllability'. Table 8.5 provides an overview of some relevant details that can be used to evaluate 'spatio-temporal controllability' in these three categories. For the approval process, further detailed information can be added and combined in these criteria e.g. the number of organisms to be released, specific regional biodiversity, abundance of protected species, occurrence of plant and animal pests and other relevant data, if available.

Case Studies: How to Apply 'Spatio-Temporal Controllability' in Practice

The following sections describe how 'spatio-temporal controllability' might ideally be applied in ERA. This includes two case studies for GE crops (maize and oilseed rape) and a case study on insects (olive flies, *Bactrocera oleae*), with and without gene drives. The case studies should be seen as mostly hypothetical; it is assumed that under real conditions more detailed data would be available.

Table 8.5 Some specific issues relevant for the assessment of 'spatio-temporal controllability' (vertical reading)

Biology of the target species (no GE)	Interactions of the target species with the environment (no GE)	Intended trait (GE)
Potential to persist and propagate	Role and function in food web	Is the GE organism intended to produce more than one generation after release?
Population dynamics and life cycle	Interaction with closely associated organisms (microbiome, parasites, symbiotic organisms)	How can genetic stability be controlled in following generations after the release?
Potential to spread beyond fields/into different ecosystems	Interaction with useful species and the wider environment (beneficial insects, soil organisms, protected species)	Does the trait impact the fitness of the organisms?
Potential for gene flow and reproduction with wild populations of the target species	Role and function in energy- and nutrient-cycle	Does the trait impact the composition of biologically active compounds?
Genetic diversity in wild populations of the target species	Impact of biotic stressors e.g. pests and pathogens (whole life cycle)	Can the persistence of the organisms be determined if necessary?
Potential for gene flow to other species	Occurrence of abiotic stressors such as climate conditions (whole life cycle)	

GE Maize for Commercial Cultivation in Sweden

In the first case study, maize plants are engineered to produce a higher biomass (e.g. MON87403, see EFSA 2018) and a hypothetical application for their cultivation in Sweden has to be assessed. The outcome of the 'spatio-temporal controllability' assessment in this case (Table 8.6) is that the approval process could proceed and a full and detailed risk assessment should be conducted before a decision is taken on the safety of the crops.

GE Oilseed Rape for Commercial Cultivation in the EU

In this case, oilseed rape plants are engineered to be resistant to glyphosate (such as MON88302) and intended for release in the EU (for further references see EFSA 2014; Testbiotech 2014). The result of the 'spatio-temporal controllability' assessment (Table 8.7) is that the approval process can only proceed for releases on very limited scale, within clearly defined areas and when gene flow is prevented. Under these conditions, a full and detailed risk assessment should be conducted before

Table 8.6 Example of 'spatio-temporal controllability' assessment regarding a hypothetical application for commercial cultivation in Sweden of transgenic maize that produces higher biomass (vertical reading)

Biology of the target species	Interactions with the environment	Intended trait
Conditions in Sweden leave hardly any potential for maize plants to persist for longer periods of time	Interactions with the environment should primarily be considered in the context of the intended trait	The GE maize is not intended to produce offspring, it can be controlled for genetic stability before sowing and its occurrence in the environment can be terminated if required
There is no potential for maize to spread beyond fields into other habitats in Sweden		Metabolic pathways which interfere with plant growth are multifunctional and complex. They are connected to plant characteristics such as stress reactions, fitness and composition of the plant constituents. Under these circumstances, risk assessment should be driven by the hypothesis that the biological characteristics of the plants as a whole will be changed by the genomic intervention. This needs to be checked carefully within ERA
Gene flow to other species would not be expected in Sweden		Within ERA, the GE plants should be exposed to a broad range of stressors and different combination thereof to investigate if the biological characteristics of the plants (especially their fitness) might change due to genome × environmental interactions

Further references: EFSA (2018), Testbiotech (2018b)

a decision is taken on the safety of the crops. However, applications for releases without such control mechanisms could not proceed and would be terminated after assessment of 'spatio-temporal controllability'.

Olive Flies with RIDL for Experimental Release in Spain

In this case study, the olive flies are genetically engineered with so-called RIDL-technology ("release of insects carrying a dominant lethal genetic system") developed

Table 8.7 Example of application of 'spatio-temporal controllability' criterion in a hypothetical case in the EU of commercial cultivation of transgenic oilseed rape resistant to glyphosate (vertical reading)

Biology of the target species	Interactions with the environment	Intended trait
Oilseed rape can persist and propagate in the environment, cross with other oilseed rape plants. Seeds can remain viable in the soil over more than ten years (seed dormancy)	Kernels are taken up by wildlife species, and seeds can be transported over larger areas	Although not intended by the trait, the GE plant will produce more than one generation if it is allowed to flower and produce kernels
The plants can spread beyond agricultural fields, especially in rural habitats. Pollen can be distributed over several kilometres	The flowering plants are an important food source for bees, thereby pollen can get transported over larger areas	The insertion of the gene that renders resistance to glyphosate is assumed to unintentionally enhance fitness also in glyphosate free environment (see 2.2)
Gene flow to populations of wild relative species can occur		There are indications that the fitness of the plants is especially enhanced under stressful conditions (see 2.2)

Further references: EFSA (2014), Testbiotech (2014)

by Oxitec (Ant et al. 2012). The effects are gender-specific: male transgenic flies will mate with the native female flies and thereby introduce their artificial genes into the native populations. While the male offspring will survive, the female offspring will die at the larval stage. As a result, the natural population of olive flies will supposedly decrease (for more background see Ant et al. 2012).

The outcome of the 'spatio-temporal controllability' assessment (Table 8.8) is that the approval process can only proceed if the olive flies are kept in cages and gene flow to native populations is prevented. However, sufficiently robust risk assessment could not be carried out if gene flow to natural populations were to occur. Therefore, an approval process for any release without high-level safety caging could not proceed and would be terminated. The 'spatio-temporal controllability' assessment might possibly yield other results if the field trials were located in areas where no olive flies occur naturally.

Experiments with Gene Drive in Olive Flies

In this case, it is assumed that the olive flies would be genetically engineered with a gene drive that would cause female offspring to die and leave the male offspring to survive and spread. Possible approaches are described by Champer et al. (2017).

Most criteria presented in Table 8.8 are also relevant for gene drives. However, in regard to the intended trait, the spread of the genes within natural populations would

Table 8.8 Example of 'spatio-temporal controllability' assessment for hypothetical experimental field trials of GE olive flies in Spain (vertical reading)

Biology of the target species	Interactions with the environment	Intended trait
Olive flies are a wild species that can persist and propagate in the whole of the Mediterranean area, and in regions with a similar climate. Their habitat is not clearly confined, except for the presence of olive trees (Daane and Johnson 2010)	There are complex interactions with other species such as birds, spiders, ants, chalcid wasps and symbiotic bacteria (Bigler et al. 1986; Daane and Johnson 2010; Gonçalves et al. 2012; Neuenschwander et al. 1983; Picchi et al. 2016)	Once released, the GE flies will mate in natural populations and cause the emergence of next generations without human intervention. Next generation effects might occur without being noticed
Under specific conditions, such as high population densities, maximum dispersal distances for olive flies reported in literature range from 4000 to 5000 m (Economopoulos et al. 1978; Remund et al. 1976)	The interrelationships include grazing, predation and symbiosis. The interrelations vary greatly throughout the fly's life history and different developmental stages (egg, larva, pupa, adult)	
Population dynamics and life cycle go through several stages (egg, larva, pupa, adult) and are subjected to winter seasons, creating potential bottlenecks in regional populations (Augustinos et al. 2005; Ochando and Reyes 2000)	There are specific and symbiotic microbes associated with the olive flies (Ben-Yosef et al. 2014; Capuzzo 2005)	The trait is unlikely to enhance fitness
Molecular analyses indicate a high level of gene flow among the Mediterranean populations (Augustinos et al. 2005; Ochando and Reyes 2000; Segura et al. 2008)		It can be assumed that, depending on the amount and frequency of GE flies released, they might be eliminated by natural processes after a period of time (Preu et al. 2019). However, various factors can have an impact on these processes and their actual duration cannot be determined
There are other known species that can mate with olive flies, however, it is unclear whether they can produce viable offspring and enable gene flow (Schutze et al. 2013)		

exceed the Mendelian pattern of inheritance. Consequently, the artificial genes would spread rapidly within the natural populations and the elimination of the GE flies via natural processes is less likely, or would at least take longer compared to the case study in Table 8.8.

Therefore, the outcome of the 'spatio-temporal controllability' assessment is that the approval process should only proceed if the olive flies are kept in the laboratory in regions where no native populations of olive flies occur. However, approval processes for experiments in regions where these flies occur naturally (such as the Mediterranean area) could not proceed because of lack of 'spatio-temporal controllability'. The assessment of 'spatio-temporal controllability' might come to other conclusions where the flies are kept in a laboratory with very high safety standards (for further reasoning see: Testbiotech 2018b).

The Role of the Risk Manager

In the EU, the regulatory system for GE organisms is based on a system of risk analysis set out in Regulation 178/2002: risk analysis is based on risk assessment (carried out by the EFSA) and risk management (carried out by the EU Commission and the member states). Additional regulations concern specific aspects such as environmental releases (Dir. 2001/18) and food and feed safety (Regulation 1829/2003).

In all decision making in the EU, approval process lies with the risk manager and in risk assessment policy (JRC 2008). Therefore, the EU Commission and the EU member states have to make sure that their decisions are based on a sufficiently robust risk assessment. Reliable decision making as required in EU regulation is not possible without reliable risk assessment: products can only be released or allowed on the market if they are shown to be safe (EU Directive 2001/18; EU Regulation 1829/2003). Without reliable risk assessment, no reliable decision-making as requested by EU regulations is possible (see Fig. 8.2).

In this context, the risk manager, and especially the EU Commission, can make use of their power to set adequate standards in risk assessment by establishing a robust framework for the EFSA (JRC 2008). It is interesting to note that the EU Commission adopted Regulation 503/2013 which sets the standards for assessing food and feed safety. However, no similar regulation has as yet been set by the EU Commission in regard to environmental risk assessment.

In regard to spatio-temporal control, EU Directive 2001/18 could be used as a legal basis to set the relevant standards: according to Krämer (2013), spatio-temporal control is a necessary prerequisite to enable the PP. He comes to the conclusion that "Where there is, in a concrete case, a likelihood that genetically modified plants or animals cannot be retrieved, the legal obligation to ensure that any release must be 'safe' requires the refusal to authorize such releases." (Paragraph 250) However, Krämer also shows that there are significant uncertainties in the implementation of EU regulation that require further attention and which could be ruled out by additional implementing regulations. At a certain point within such a framework, the

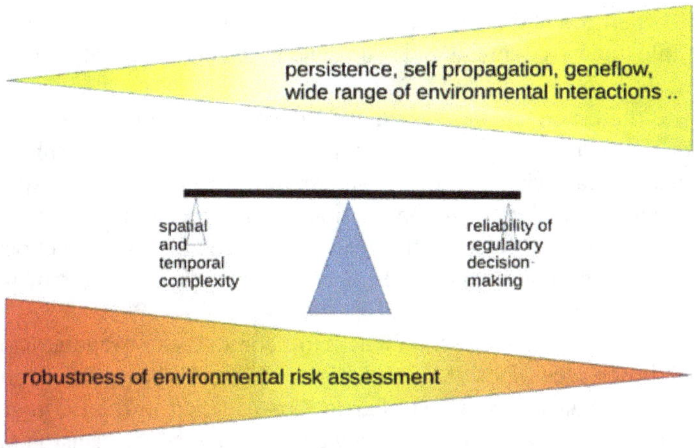

Fig. 8.2 The balance between spatio-temporal complexity and the reliability of risk management decision making

EU Commission could request EFSA to assess 'spatio-temporal controllability' to deal with substantial uncertainties and non-knowledge.

As addressed in the case of the olive fly, the risk assessment of an application for laboratory uses or experimental releases of GE organisms can also become a challenge for national regulatory authorities. Thus, national legislation should foresee adequate regulatory oversight and request 'spatio-temporal controllability' to be applied as a step within the approval process.

There are further reasons why the risk manager should use 'spatio-temporal controllability' to determine relevant cut-off criteria:

- Spontaneous transboundary movements: if GE organisms can spontaneously cross borders, their release can be considered to be a violation of rights under the so-called Cartagena Protocol (CBD 2000).
- No possibility of coexistence: if coexistence with relevant standards for food production, such as organic agriculture, is not possible, the release of GE organisms would infringe consumers' choice and the livelihoods of organic farmers (Reeves and Phillipson 2017).

Discussion

The development of gene drive organisms and other GE organisms that can persist and self-replicate in the environment and/or can cross with natural populations create new challenges in risk assessment. As described, GE organisms inheriting a gene drive differ substantially in their characteristics compared to other GE organisms assessed by EFSA and other regulatory authorities. Existing EFSA guidance shows

(EFSA 2013) that the authority is already well aware that risk assessment of GE organisms able to persist, spread and propagate in the environment will bring about new challenges.

As shown, these new challenges chiefly concern next generation effects: there are many reasons why the spontaneous offspring of GE organisms might differ in their biological characteristics from one generation to the next, especially if the gene constructs are introduced into wild populations. Closely related to this challenge are questions concerning the impact of the genetic background of the target populations and interactions with abiotic and biotic stressors, as well as aspects of population dynamics and life cycle.

The basic challenge for risk assessment in this context is how regulatory decisions can be made in the face of substantial non-knowledge. To solve this problem, it is proposed to apply cut-off criteria similar to those applied in the EU regulation of chemicals. To define these cut-off criteria within the regulatory decision making on GE organisms, a new step in the risk assessment of GE organisms should be applied, i.e. 'spatio-temporal controllability'. This step is composed of three criteria: (1) the biology of the target organism, (2) its known interaction with the environment and (3) the biological characteristics of the GE organisms.

This approach uses specific 'knowns' to decide upon 'known unknowns' (such as next generation effects and genomic × environmental interactions). It is assumed, the criterion of 'spatio-temporal controllability' can inform regulatory decision making even in the light of major uncertainties emerging from the spatio-temporal dimension. This can be seen as the equivalent of cut-off criteria such as "PBT" and "vPvB" that are anchored in the EU regulation of chemical substances.

It should, however, be recognised that the assessment of 'spatio-temporal controllability' is just a step within risk assessment and not a replacement for it. In this context, it is also important to acknowledge that there are further issues, such as horizontal gene transfer, which concern the spatio-temporal control of GE organisms. Therefore, environmental risk assessment cannot and should not be reduced to the step of 'spatio-temporal controllability'. Risk assessment might well be terminated after 'spatio-temporal controllability' assessment, but if it proceeds, all other steps and criteria still have to be applied.

This approach is exemplified in case studies for the cultivation of GE maize (in Sweden) and oilseed rape (in the EU), as well as experimental releases of olive flies, with and without gene drives (in Spain). Preliminary results show that the assessment of 'spatio-temporal controllability' produces results which are meaningful and allow the application of cut-off criteria within the process of risk assessment: the authorisation process should not proceed and the release of the GE organism should not be allowed if it is known that the GE organisms are able to escape 'spatio-temporal controllability' due to propagation in natural populations, with no effective control of spread or way of preventing persistence in the environment.

The schematic and partially hypothetical cases as presented lack some data and information that under real conditions would be included in the pending application dossier. Further, in practice the results would depend on a shifting baseline of information and might therefore differ from future results. The approach as proposed can be considered flexible enough to be improved by adding further criteria and new data. It can be applied very generally to applications for environmental releases, no matter if these concern experimental field trials or commercial cultivation.

As far as the role of the risk manager in the EU is concerned, it should be acknowledged that applications for releases of GE organisms that lack spatio-temporal control can already be rejected, even without an additional and specific step in risk assessment as suggested. However, especially in the light of the PP, the application of cut-off criteria within the process of risk assessment has some significant, important and convincing advantages: it can provide more clarity, transparency and also more reliability in final decision making. Further, it can save on resources, since this additional step in risk assessment would influence the approval process at an early stage.

Finally, these criteria can be used to inform upstream processes and thereby generate more clarity and certainty at an early stage of research and development. Many researchers currently developing gene drive applications are already aware of the problem of spatio-temporal complexity (see, for example, Noble et al. 2017). At present, several projects are looking to develop gene drives that can be refined to specific regions or defined periods of time (see for example Min et al. 2017). It is assumed, there is no general obstacle to the future application of 'spatio-temporal controllability' assessment for these developments, and therefore meaningful results could be expected.

Conclusions

New challenges arise with applications such as 'gene drive' that can be introduced into natural populations where they can propagate and spread further. Due to the complexity of the biology of these organisms and their interactions with the environment, increasing uncertainty and areas of non-knowledge have to be taken into account.

It has been shown that risk assessment of intended environmental releases of GE organisms linked to self-propagation of artificial genetic elements over several generations will suffer from major uncertainties and unknowns, emerging in most cases from next generation effects. It can be assumed that a tipping point can emerge at a certain point in the dissolution of spatio-temporal boundaries where it becomes necessary to apply cut-off criteria and stop the approval process. This means that risk assessors and risk managers face the problem of how to come to robust conclusions and reliable decisions within the approval process that also give substantial weight to the PP.

It is proposed to introduce cut-off criteria, based on a specific step of 'spatio-temporal controllability' within risk assessment. This new step combines three criteria:

(1) the biology of the target organisms,
(2) their naturally occurring interactions with the environment (biotic and abiotic),
(3) the intended biological characteristics (trait) of the GE organisms.

The combination of these three criteria in one specific, additional step in risk assessment has the advantage of them already being used to some extent in current EFSA risk assessment; many of the details to assess these criteria are also very well known. If it is known that GE organisms can escape 'spatio-temporal controllability' because they can propagate within natural populations with no effective control of spread or persistence, then the authorisation process cannot proceed and the release of the GE organism cannot be allowed. This concept can be used to delineate some of the boundaries between known and unknowns considered to be crucial. Further, it can help to develop an adequate regime for risk assessment which overcomes problems with the so-called comparative approach (EFSA 2010). This will foster the robustness of risk assessment and can substantially benefit the reliability of decision making within approval processes.

References

Adamczyk, J. J., Perera, O., & Meredith, W. R. (2009). Production of mRNA from the cry1Ac transgene differs among Bollgard® lines which correlates to the level of subsequent protein. *Transgenic Research, 18*(1), 143–149. https://doi.org/10.1007/s11248-008-9198-z.

Adamczyk, J. J. J., & Meredith, W. R. J. (2004). Genetic basis for variability of Cry1Ac expression among commercial transgenic *Bacillus thuringiensis* (Bt) cotton cultivars in the United States. *The Journal of Cotton Science, 8*(1), 433. https://pubag.nal.usda.gov/catalog/10670.

Akbari, O. S., Chen, C.-H., Marshall, J. M., Huang, H., Antoshechkin, I., & Hay, B. A. (2014). Novel synthetic medea selfish genetic elements drive population replacement in Drosophila, and a theoretical exploration of medea-dependent population suppression. *ACS Synthetic Biology, 3*(12), 915–928. https://doi.org/10.1021/sb300079h.

Ant, T., Koukidou, M., Rempoulakis, P., Gong, H.-F., Economopoulos, A., Vontas, J., et al. (2012). Control of the olive fruit fly using genetics-enhanced sterile insect technique. *BMC Biology, 10*, 51. https://doi.org/10.1186/1741-7007-10-51.

Augustinos, A. A., Mamuris, Z., Stratikopoulos, E. E., D'Amelio, S., Zacharopoulou, A., & Mathiopoulos, K. D. (2005). Microsatellite analysis of olive fly populations in the Mediterranean indicates a westward expansion of the species. *Genetica, 125*(2–3), 231–241. https://doi.org/10.1007/s10709-005-8692-y.

Bauer-Panskus, A., Breckling, B., & Hamberger, S., Then, C. (2013). Cultivation-independent establishment of genetically engineered plants in natural populations: Current evidence and implications for EU regulation. *Environmental Sciences Europe 25*, 34. www.enveurope.com/content/25/1/34.

Ben-Yosef, M., Pasternak, Z., Jurkevitch, E., & Yuval, B. (2014). Symbiotic bacteria enable olive flies (*Bactrocera oleae*) to exploit intractable sources of nitrogen. *Journal of Evolutionary Biology, 27*(12), 2695–2705. https://doi.org/10.1111/jeb.12527.

Bigler, F., Neuenschwander, P., Delucchi, V., & Michelakis, S. (1986). Natural enemies of preimaginal stages of Dacus oleae Gmel. (Dipt., Tephritidae) in Western Crete. II. Impact on olive fly populations. *Bollettino del Laboratorio di Entomologia Agraria "Filippo Silvestri", Italy, 43*, 79–96.

Böschen, S. (2009). Hybrid regimes of knowledge? Challenges for constructing scientific evidence in the context of the GMO-debate. *Environmental Science and Pollution Research, 16*(5), 508–520. https://doi.org/10.1007/s11356-009-0164-y.

Böschen, S., Kastenhofer, K., Marschall, L., Rust, I., Soentgen, J., & Wehling, P. (2006). Scientific cultures of non-knowledge in the controversy over genetically modified organisms (GMO): The cases of molecular biology and ecology. *GAIA-Ecological Perspectives for Science and Society, 15*(4), 294–301(8). https://doi.org/10.14512/gaia.15.4.12.

Bollinedi, H., Gopala Krishnan, K., Prabhu, K. V., Singh, N. K., Mishra, S., Khurana, J. P., et al. (2017). Molecular and functional characterization of GR2-R1 event based backcross derived lines of golden rice in the genetic background of a mega rice variety Swarna. *PLoS ONE, 12*(1), e0169600. https://doi.org/10.1371/journal.pone.0169600.

Breckling, B. (2013). Transgenic evolution and ecology are proceeding. In B. Breckling & R. Verhoeven (Eds.), *GM-crop cultivation—Ecological effects on a landscape scale* (pp. 130–135). Frankfurt: Peter Lang.

Buchman, A., Marshall, J. M., Ostrovski, D., Yang, T., & Akbari, O. S. (2018). Synthetically engineered *Medea* gene drive system in the worldwide crop pest *Drosophila suzukii*. *Proceedings of the National Academy of Sciences of the United States of America, 115*(18), 4725–4730. https://doi.org/10.1073/pnas.1713139115.

Burgos, N. R., Singh, V., Tseng, T. M., Black, H., Young, N. D., Huang, Z., et al. (2014). The impact of herbicide-resistant rice technology on phenotypic diversity and population structure of United States weedy rice. *Plant Physiology, 166*, 1208–1220. https://doi.org/10.1104/pp.114.242719.

Cao, Q.-J., Xia, H., Yang, X., & Lu, B.-R. (2009). Performance of hybrids between weedy rice and insect-resistant transgenic rice under field experiments: Implication for environmental biosafety assessment. *Journal of Integrative Plant Biology, 51*(12), 1138–1148. https://doi.org/10.1111/j.1744-7909.2009.00877.x.

Capuzzo, C. (2005). "Candidatus Erwinia dacicola", a coevolved symbiotic bacterium of the olive fly *Bactrocera oleae* (Gmelin). *International Journal of Systematic and Evolutionary Microbiology, 55*, 1641–1647. https://doi.org/10.1099/ijs.0.63653-0.

CBD. (2000). Cartagena protocol on biosafety to the convention on biological diversity, Montreal, Secretariat of the Convention on Biological Diversity, ISBN: 92-807-1924-6.

Champer, J., Reeves, R., Oh, S. Y., Liu, C., Liu, J., Clark, A. G., et al. (2017). Novel CRISPR/Cas9 gene drive constructs in *Drosophila* reveal insights into mechanisms of resistance allele formation and drive efficiency in genetically diverse populations. bioRxiv. https://doi.org/10.1101/112011.

Chandler, C. H., Chari, S., & Dworkin, I. (2013). Does your gene need a background check? How genetic background impacts the analysis of mutations, genes, and evolution. *Trends in Genetics, 29*(6), 358–366. https://doi.org/10.1016/j.tig.2013.01.009.

Daane, K. M., & Johnson, M. W. (2010). Olive fruit fly: Managing an ancient pest in modern times. *Annual Review of Entomology, 55*, 151–169. https://doi.org/10.1146/annurev.ento.54.110807.090553.

da Silva, K. J., de Armas, R. D., Soares, C. R. F., & Ogliari, J. B. (2016). Communities of endophytic microorganisms in different developmental stages from a local variety as well as transgenic and conventional isogenic hybrids of maize. *World Journal of Microbiology and Biotechnology, 32*, 189. https://doi.org/10.1007/s11274-016-2149-6.

DiCarlo, J. E., Chavez, A., Dietz, S. L., Esvelt, K. M., & Church, G. M. (2015). Safeguarding CRISPR–Cas9 gene drives in yeast. *Nature Biotechnology, 33*, 1250–1255. https://doi.org/10.1038/nbt.3412.

Economopoulos, A. P., Haniotakis, G. E., Mathioudis, J., & Missis, N. (1978). Long-distance flight of wild and artificially-reared *Dacus oleae* (Gmelin) (Diptera, Tephritidae). *Journal of Applied Entomology, 87*(1–4), 101–108.

EFSA. (2010). Guidance on the environmental risk assessment of genetically modified plants. *EFSA Journal, 8*(11), 1879. https://doi.org/10.2903/j.efsa.2010.1879.

EFSA. (2013). Guidance on the environmental risk assessment of genetically modified animals. *EFSA Journal, 11*(5), 3200, 190 pp. https://doi.org/10.2903/j.efsa.2013.3200.

EFSA. (2014). Scientific opinion on application (EFSAGMO-BE-2011–101) for the placing on the market of herbicide-tolerant genetically modified oilseed rape MON 88302 for food and feed uses, import and processing under Regulation (EC) No.1829/2003 from Monsanto. *EFSA Journal, 12*(6), 3701, 37 pp. https://doi.org/10.2903/j.efsa.2014.3701.

EFSA. (2018). Scientific Opinion on the assessment of genetically modified maize MON 87403 for food and feed uses, import and processing, under Regulation (EC) No 1829/2003 (application EFSA-GMO-BE-2015–125). *EFSA Journal, 16*(3), 5225, 28 pp. https://doi.org/10.2903/j.efsa.2018.5225.

Evangelou, A., Ignatiou, A., Antoniou, C., Kalanidou, S.,Hadjimatheou, S., Ellina S., et al. (2018). Unpredictable effects of the genetic background of transgenic lines in physiological quantitative traits, bioRxiv preprint, https://dx.doi.org/10.1101/494419.

Fang, J., Nan, P., Gu, Z., Ge, X., Feng, Y.-Q., & Lu, B.-R. (2018). Overexpressing exogenous 5-enolpyruvylshikimate-3-phosphate synthase (EPSPS) genes increases fecundity and auxin content of transgenic Arabidopsis plants. *Frontiers in Plant Science, 9*. https://doi.org/10.3389/fpls.2018.00233.

Gantz, V. M., & Bier, E. (2015). The mutagenic chain reaction: A method for converting heterozygous to homozygous mutations. *Science, 348*(6233), 442–444. https://science.sciencemag.org/content/348/6233/442v.

Gonçalves, F. M., Rodrigues, M. C., Pereira, J. A., Thistlewood, H., & Torres, L. M. (2012). Natural mortality of immature stages of *Bactrocera oleae* (Diptera: Tephritidae) in traditional olive groves from north-eastern Portugal. *Biocontrol Science and Technology, 22*, 837–854. https://doi.org/10.1080/09583157.2012.691959.

Grunwald, H. A., Gantz, V. M., Poplawski, G., Xu, X. S., Bier, E., & Cooper, K. L. (2018). Super-Mendelian inheritance mediated by CRISPR/Cas9 in the female mouse germline. bioRxiv. https://doi.org/10.1101/362558

Hammond, A., Galizi, R., Kyrou, K., Simoni, A., Siniscalchi, C., Katsanos, D., et al. (2015). CRISPR-Cas9 gene drive system targeting female reproduction in the malaria mosquito vector *Anopheles gambiae*. *Nature Biotechnology, 34*, 78–83. https://www.nature.com/articles/nbt.3439.

Hilbeck, A., & Otto, M., 2015. Specificity and combinatorial effects of *Bacillus thuringiensis* Cry toxins in the context of GMO risk assessment. *Frontiers Environmental Science, 3*, 71. https://doi.org/10.3389/fenvs.2015.00071.

JRC. (2008). Risk-assessment policies: Differences across jurisdictions, Joint Research Centre—Institute for Prospective Technological Studies, European Commission, Authors: Erik Millstone, Patrick van Zwanenberg, Les Levidow, Armin Spök, Hideyuki Hirakawa, Makiko Matsuo, Scientific and Technical Research series—ISSN 1018-5593. https://ec.europa.eu/jrc/en/publication/eur-scientific-and-technical-research-reports/risk-assessment-policies-differences-across-jurisdictions.

KaramiNejadRanjbar, M., Eckermann, K. N., Ahmed, H. M. M., Sánchez, C. H. M., Dippel, S., Marshall, J. M., et al. (2018). Consequences of resistance evolution in a Cas9-based sex conversion-suppression gene drive for insect pest management. *Proceedings of the National Academy of Sciences, 115*(24), 6189–6194. https://doi.org/10.1073/pnas.1713825115.

Katsuta, K., Matsuo, K., Yoshimura, Y., & Ohsawa, R. (2015). Long-term monitoring of feral genetically modified herbicide-tolerant *Brassica napus* populations around unloading Japanese ports. *Breeding Science, 65*(3), 265–275. https://doi.org/10.1270/jsbbs.65.265.

Kawata, M., Murakami, K., Ishikawa, T. (2009). Dispersal and persistence of genetically modified oilseed rape around Japanese harbors. *Environmental Science and Pollution Research, 16*(2), 120–126. https://doi.org/10.1007/s11356-008-0074-4.

Knispel, A. L., & McLachlan, S. M. (2010). Landscape-scale distribution and persistence of genetically modified oilseed rape (*Brassica napus*) in Manitoba, Canada. *Environmental Science and Pollution Research, 17*, 13–25. https://doi.org/10.1007/s11356-009-0219-0.

Kuzma, J., Gould, F., Brown, Z., Collins, J., Delborne, J., Frow, E., et al. (2017). A roadmap for gene drives: Using institutional analysis and development to frame research needs and governance in a systems context. *Journal of Responsible Innovation, 5*, S13–S39. https://doi.org/10.1080/23299460.2017.1410344.

Kyrou, K., Hammond, A. M., Galizi, R., Kranjc, N., Burt, A., Beaghton, A. K., et al. (2018). A CRISPR–Cas9 gene drive targeting doublesex causes complete population suppression in caged Anopheles gambiae mosquitoes. *Nature Biotechnology, 36*, 1062–1066. https://doi.org/10.1038/nbt.4245.

Krämer, L. (2013). Genetically modified living organisms and the precautionary principle. www.testbiotech.de/node/904.

Ledford, H. (2015). CRISPR, the disruptor. *Nature, 522*(7554), 20–24. www.nature.com/news/crispr-the-disruptor-1.17673.

Lynch, S. V., & Pedersen, O. (2016). The human intestinal microbiome in health and disease. *New England Journal of Medicine, 375*(24), 2369–2379. https://doi.org/10.1056/NEJMra1600266.

Lu, B.-R., & Yang, C. (2009). Gene flow from genetically modified rice to its wild relatives: Assessing potential ecological consequences. *Biotechnology Advances, Biotechnology for the Sustainability of Human Society Invited Papers from IBS,* 27(6), 1083–1091. https://doi.org/10.1016/j.biotechadv.2009.05.018.

Matthews, D., Jones, H., Gans, P., Coates, S., & Smith, L. M. J. (2005). Toxic secondary metabolite production in genetically modified potatoes in response to stress. *Journal of Agriculture and Food Chemistry, 53*(20), 7766–7776. https://doi.org/10.1021/jf050589r.

Meyer, P., Linn, F., Heidmann, I., Meyer, H., Niedenhof, I., & Saedler, H. (1992). Endogenous and environmental factors influence 35S promoter methylation of a maize A1 gene construct in transgenic petunia and its colour phenotype. *Molecular and General Genetics MGG, 231*(3), 345–352. https://doi.org/10.1007/BF00292701.

Min J., Noble C., Najjar D., & Esvelt K. M. (2017). Daisyfield gene drive systems harness repeated genomic elements as a generational clock to limit spread, bioRxiv preprint first posted online February 6, 2017. https://doi.org/10.1101/104877.

Mullis, M. N., Matsui, T., Schell, R., Foree, R., & Ehrenreich, I. M. (2018). The complex underpinnings of genetic background effects. *Nature Communications, 9*, 3548. https://doi.org/10.1038/s41467-018-06023-5.

Nardi, J., Moras, P. B., Koeppe, C., Dallegrave, E., Leal, M. B., & Rossato-Grando, L. G. (2017). Prepubertal subchronic exposure to soy milk and glyphosate leads to endocrine disruption. *Food and Chemical Toxicology, 100*, 247–252. https://doi.org/10.1016/j.fct.2016.12.030.

Neuenschwander, P., Bigler, F., Delucchi, V., & Michelakis, S. (1983). Natural enemies of preimaginal stages of *Dacus oleae* Gmel. (Dipt., Tephritidae) in Western Crete. I. Bionomics and phenologies. *Bollettino del Laboratorio di Entomologia Agraria Filippo Silvestri, Portici, 40*, 3–32.

Noble C., Adlam B., Church G. M., Esvelt K. M., & Nowak M. A. (2017). Current CRISPR gene drive systems are likely to be highly invasive in wild populations. *eLife, 7*, e33423. https://doi.org/10.7554/eLife.33423.

Ochando, M. D., & Reyes, A. (2000). Genetic population structure in olive fly *Bactrocera oleae* (Gmelin): Gene flow and patterns of geographic differentiation. *Journal of Applied Entomology, 124*(3–4), 177–183. https://doi.org/10.1046/j.1439-0418.2000.00460.x.

Oladosu, Y., Rafii, M. Y., Abdullah, N., Hussin, G., Ramli, A., Rahim, H. A., et al. (2016). Principle and application of plant mutagenesis in crop improvement: A review. *Biotechnology & Biotechnological Equipment, 30*, 1–16. https://doi.org/10.1080/13102818.2015.1087333.

Oye, K. A., Esvelt, K., Appleton, E., Catteruccia, F., Church, G., Kuiken, T., et al. (2014). Regulating gene drives. *Science, 345*(6197), 626–628. https://doi.org/10.1126/science.1254287.

Picchi, M. S., Bocci, G., Petacchi, R., & Entling, M. H. (2016). Effects of local and landscape factors on spiders and olive fruit flies. *Agriculture, Ecosystems & Environment, 222*, 138–147. https://doi.org/10.1016/j.agee.2016.01.045.

Preu, M., Breckling, B., & Schröder, W. (2019). Case study 1: Olive fruit fly, Bactrocera oleae, GeneTip Genetic innovations as a trigger for phase transitions in the population dynamics of animals and plants. www.genetip.de/en/biotip-pilot-study/ (unpublished).

Reeves, R. G., & Phillipson, M. (2017). Mass releases of genetically modified insects in area-wide pest control programs and their impact on organic farmers. *Sustainability, 9*(1), 59. https://doi.org/10.3390/su9010059.

Remund, U., Boller, E. F., Economopoulos, A. P., & Tsitsipis, J. A. (1976). Flight performance of *Dacus oleae* reared on olives and artificial diet. *Zeitschrift für Angewandte Entomologie, 82*(1–4), 330–339. https://doi.org/10.1111/j.1439-0418.1976.tb03420.x.

Richardson, L. A. (2017). Evolving as a holobiont. *PLoS Biology, 15*(2), e2002168. https://doi.org/10.1371/journal.pbio.2002168.

Saltz, J. B., Bell, A. M., Flint, J., Gomulkiewicz, R., Hughes, K. A., & Keagy, J. (2018). Why does the magnitude of genotype-by-environment interaction vary? *Ecology and Evolution, 8*(12), 6342–6353. https://doi.org/10.1002/ece3.4128.

Schaefer, H. M., & Ruxton, G. D. (2011). Plant-animal communication. Oxford: OUP. ISBN: 978-0-19-162097-3

Schutze, M. K., Jessup, A., Ul-Haq, I., Vreysen, M. J. B., Wornoayporn, V., Vera, M. T., et al. (2013). Mating compatibility among four pest members of the *Bactrocera dorsalis* fruit fly species complex (Diptera: Tephritidae). *Journal of Economic Entomology, 106*(2), 695–707. https://doi.org/10.1603/EC12409.

Segura, M. D., Callejas, C., & Ochando, M. D. (2008). *Bactrocera oleae*: A single large population in Northern Mediterranean basin. *Journal of Applied Entomology, 132*(9–10), 706–713. https://doi.org/10.1111/j.1439-0418.2008.01366.x.

Simon, S., Otto M., & Engelhard M. (2018). Synthetic gene drive: Between continuity and novelty: Crucial differences between gene drive and genetically modified organisms require an adapted risk assessment for their use. *EMBO Reports, 19*(5), e45760. https://doi.org/10.15252/embr.201845760.

Testbiotech. (2014). Testbiotech comment on EFSA GMO Panel Scientific Opinion on application (EFSAGMO-BE-2011–101) for the placing on the market of herbicide-tolerant genetically modified oilseed rape MON 88302 for food and feed uses, import and processing under Regulation (EC) No 1829/2003 from Monsanto, www.testbiotech.org/sites/default/files/TBT%20comment_oilseed_rape_MON_88302.pdf.

Testbiotech. (2018a). Testbiotech comment on EFSA GMO Panel, 2018, Scientific opinion on the assessment of genetically engineered maize MON 87403 for food and feed uses, import and processing, under Regulation (EC) No 1829/2003 (application EFSA-GMO-BE-2015-s125). https://www.testbiotech.org/sites/default/files/Testbiotech%20comment_maize_MON87403.pdf.

Testbiotech. (2018b). Kommentar zum "Entwurf einer Verordnung über die Sicherheitsstufen und Sicherheitsmaßnahmen bei gentechnischen Arbeiten in gentechnischen Anlagen (Gentechnik-Sicherheitsverordnung – GenTSV)". www.testbiotech.org/node/2228.

Then, C., & Lorch, A. (2008). A simple question in a complex environment: How much Bt toxin do genetically engineered MON810 maize plants actually produce? In: *Implications of GM-crop cultivation at large spatial scales*. Theorie in der Ökologie (pp. 17–21). Frankfurt: Peter Lang.

Trtikova, M., Wikmark, O. G., Zemp, N., Widmer, A., & Hilbeck, A. (2015). Transgene expression and Bt protein content in transgenic Bt maize (MON810) under optimal and stressful environmental conditions. *PLoS ONE, 10*(4), e0123011. https://doi.org/10.1371/journal.pone.0123011.

Vacher, C., Weis, A. E., Hermann, D., Kossler, T., Young, C., & Hochberg, M. E. (2004). Impact of ecological factors on the initial invasion of Bt transgenes into wild populations of birdseed rape (*Brassica rapa*). *TAG. Theoretical and Applied Genetics, 109*(4), 806–814. https://doi.org/10.1007/s00122-004-1696-7.

Warwick, S. I., Légère, A., Simard, M.-J., & James, T. (2008). Do escaped transgenes persist in nature? The case of an herbicide resistance transgene in a weedy *Brassica rapa* population. *Molecular Ecology, 17*(5), 1387–1395. https://doi.org/10.1111/j.1365-294X.2007.03567.x.

Wynne, B. (1992). Uncertainty and environmental learning: Reconceiving science and policy in the preventive paradigm. *Global environmental change, 2*(2), 111–127. https://doi.org/10.1016/0959-3780(92)90017-2.

Yang, X., Li, L., Jiang, X., Wang, W., Cai, X., Su, J., et al. (2017). Genetically engineered rice endogenous 5-enolpyruvoylshikimate-3-phosphate synthase (epsps) transgene alters phenology and fitness of crop-wild hybrid offspring. *Scientific Reports, 7*, 6834. https://doi.org/10.1038/s41598-017-07089-9.

Zeller, S. L., Kalininal, O., Brunner, S., Keller, B., & Schmid, B. (2010). Transgene × environment interactions in genetically modified wheat. *PLoS ONE, 5*(7), e11405. https://www.plosone.org/article/info:doi/10.1371/journal.pone.0011405.

Zhu, X., Sun, L., Kuppu, S., Hu, R., Mishra, N., Smith, J., et al. (2018). The yield difference between wild-type cotton and transgenic cotton that expresses IPT depends on when water-deficit stress is applied. *Scientific Reports, 8*, 2538. https://doi.org/10.1038/s41598-018-20944-7.

9

SPAGE Technologies and Risk Management

Arnim von Gleich

In view of the rapid dynamics of genetic engineering development (in particular regarding the 'new gene-technologies' gene editing, self-propagating artificial genetic elements (SPAGE) and synthetic biology), the question is being intensively discussed whether the currently practiced risk governance[1] of the release of genetically modified organisms is sufficient to guarantee the desired high level of health, consumer and environmental safety.[2] "In the United States, it is clear that gene drive activities will trigger a variety of governance mechanisms. However, some of these mechanisms may be inadequate for identifying immediate and long-term potential environmental and public health implications of individual gene-drive applications because they lack clarity in their jurisdiction, they are challenged by the novel characteristics of gene drives, or they provide insufficient structures for public engagement" (National Academies of Sciences 2016, p. 158). Less attention is paid to the broader question of how the precautionary principle can be more strongly integrated into the

[1]Risk governance is to be understood here as an overarching concept. It encompasses risk assessment, risk evaluation, risk management, risk communication and risk regulation, as well as activities at civil society level (environmental and consumer protection) and at company level (occupational health and safety, quality assurance) that are not necessarily triggered by government directives. "Risk governance extends to issues of institutional design, legislative procedure, consultative style, organizational culture, expert accreditation, stakeholder negotiation, conflict resolution and exercise of power" Stirling et al. (2006, p. 286).

[2]"The lack of guidance from the US. Federal government applicable to ecological risk assessment for the gene drive research community is a critical gap" National Academies of Sciences (2016, p. 119). See also Roller (2005), Oye et al. (2014, p. 6197), Caplan et al. (2015), Winter (2016), Simon et al. (2018).

A. von Gleich (✉)
Department of Technological Design and Development, Faculty Production Engineering, University of Bremen, Bremen, Germany
e-mail: gleich@uni-bremen.de

risk governance of new gene-technologies.[3] Few exceptions to this are activities in Switzerland and Austria (cf. Ammann et al. 2007; Eckerstorfer et al. 2010). Possibilities for a stronger integration of the precautionary principle into the governance of new technologies are the subject of this text.

The precautionary principle plays an important role in international, European and national regulation.[4] Its interpretation is controversial, little operationalized and the subject of intense discourse.[5] The precautionary principle legitimizes precautionary measures especially when it would be irresponsible to wait until a risk can really be proven. This waiting would be particularly irresponsible if serious and/or far-reaching hazards with tendencies towards irreversible exposures and adverse effects were involved, after the occurrence of which corrective action cannot be taken.

At least in Europe, the discussion has revealed the following combination of prerequisites for precautionary measures with regard to technological innovations[6]:

(a) Lack of knowledge (reaching from uncertainty to ignorance)
(b) Comprehensible reasons for concern (indications pointing to particularly powerful, irreversible and far-reaching technological effects or to particularly serious consequences affecting irreplaceable values, particularly vulnerable population groups or ecosystems)
(c) A rudimentary cost–benefit analysis (in which at least the expected costs of precautionary measures are compared with the expected costs of inaction, or in which, for example, medical applications for which few or no alternatives exist are given more weight than applications in the food chain in which numerous alternatives exist)
(d) The availability of adequate and proportionate measures (besides risk communication and participation ranging from labelling, certifications, accreditations and codes of conduct through containment or moratorium up to substitution by less problematic alternatives).

Although the precautionary principle is well represented in the political and legal bases for action at international and European level, its anchoring and operationalization in the technology-related regulations must so far be described as rather rudimentary. Fisher et al. write: "In particular, the messy business of integrating the principle into existing institutions and relating it to well-established decision-making processes has not received the attention it should have" (2006, p. 1). The integration of the precautionary principle into the governance of self-propagating artificial genetic

[3] In the publication of the National Academies of Sciences (2016), the current need for policy reform is at least addressed as an opportunity to broaden the view: "The novelty of this technology also provides an opportunity to reflect more generally on the principles governing scientific research and suggest areas for improvement" p. 137.

[4] See in particular UNCED 1992 Principle 15; UNEP 2000 Cartagena Protocol on Biosafety tiret 9; Treaty of the Functioning of the European Union TFEU 2007 Article 191(2).

[5] Cf. e.g. Commission of the European Communities (2000), European Environment Agency (2002), Fisher et al. (2006), Stirling (2016), European Commission (2017).

[6] Commission of the European Communities (2000), Renn et al. (2003), von Schomberg (2006), Stirling et al. (2006), Amman et al. (2007), Stirling (2016), Persson (2017).

elements (SPAGE) is thus a challenging task. This is partly due to the fact that all levels and elements of governance must be included, i.e. risk assessment and evaluation (including guidelines on best practice or methods for cost–benefit assessment[7]), risk management and risk regulation at various levels. On the other hand, the task affects all phases of the innovation cycle, from research and development through process and product approval up to post-release monitoring. And finally, a product-based, process-based or function-based approach can be taken, or all three approaches can be pursued in an integrated manner (Oye et al. 2014; Sprink et al. 2016; Ishii and Araki 2016).

Reasons for Concern as an Interface Between Risk Assessment and Risk Management

In the recent past, scientific and public debates relating to the safety of genetic engineering processes and products have increasingly focused on applications in the food chain. As a result, the European Food Safety Authority (EFSA) and its 'Panel on Genetically Modified Organisms' became particularly important.[8] Genetic engineering governance thus increasingly focused on the relatively late innovation phase of product approval. However, since gene drives have so far mainly been in the research and development phase, a "governance of science and technology" is required for these new gene-technologies (National Academies of Sciences 2016, p. 138ff). This must go far beyond the existing regulations on laboratory or facility safety and deliberate release.[9]

Efforts to anchor the precautionary principle more firmly in the governance of new genetic technologies focus on risk assessment procedures and the interface between risk assessment and risk management. The necessary reforms are dealing with the further development of existing methods of environmental risk assessment (ERA) towards a precautionary hazard and exposure assessment that also takes appropriate

---

[7]The European Commission's communication on the applicability of the precautionary principle emphasizes that cost–benefit analysis must not be a matter of purely economic cost–benefit considerations, but also of "the efficiency of possible options and their public acceptance" (p. 5).

[8]Between 2003 and 2019 there were 367 EFSA publications on genetically modified organisms https://www.efsa.europa.eu/de/publications/?f%5B0%5D=im_field_subject%3A61906, last accessed 20.02.2019.

[9]Cf. German Genetic Engineering Act with its differentiation of genetic engineering facilities for research and commercial purposes, the introduction of safety levels and the involvement of the Central Commission for Biological Safety (ZKBS), as well as the approval of genetic engineering facilities (last amendment 2017) including the Genetic Engineering Procedure Ordinance (GenTVfV) and the Genetic Engineering Protection Ordinance (GenTSV) (last amendment 2015), then the EU Directive on the contained use of genetically modified microorganisms (Directive 90/219/EEC, now Directive 2009/41/EEC), the EU Directive on the protection of workers from risks related to exposure to biological agents at work (Directive 90/679/EEC, now Directive 2000/54/EEC) and the EU Directive on the deliberate release into the environment of genetically modified organisms (Directive 90/220/EEC, now Directive 2001/18/EEC).

account of various forms of lacking knowledge, and with their more stringent linkage with precautionary risk management with the help of the construct of 'reasons for concern'. In addition, the design and establishment of systematic and orderly procedures for enforcement are necessary (administrative regulations). These should also include participation opportunities for the public and civil society actors (e.g. public hearings). Finally, those institutions that are responsible for these procedures, in which information is collected, evaluation criteria are sharpened and, to a certain extent, weighing processes are carried out, must prepare themselves for a corresponding reorientation towards a precautionary risk and exposure assessment and evaluation that starts early in the innovation process.[10]

Central to the implementation of the precautionary principle within the framework of risk assessment and its interface with risk management is the identification of those 'reasons for high concern' which are capable of triggering precautionary measures. If not only the probabilities of occurrence are unknown, but also the contours of possible threat scenarios are unclear, then it is a matter of scientifically comprehensible information indicating that particularly severe and/or far-reaching consequences must be reckoned with. The EU Commission's Communication on the applicability of the precautionary principle speaks of "reasonable grounds for concern" in this respect (Commission of the European Communities 2000, p. 3; cf. also von Schomberg 2006, p. 19). In another part of the Communication 'sufficient' or 'reasonable' grounds for concern are mentioned (pp. 11 and 31). The precautionary principle is applicable in "those specific circumstances where scientific evidence is insufficient, inconclusive or uncertain and there are indications through preliminary objective scientific evaluation that there are reasonable grounds for concern that the potentially dangerous effects on the environment, human, animal or plant health may be inconsistent with the chosen level of protection" (p. 9f). The elaboration and establishment of methods and criteria for determining such reasons for concerns against the background of uncertainty and ignorance are therefore of central importance. Thus, despite incomplete knowledge, reasons for concern that could trigger precautionary measures must indicate the possibility, severity and range of adverse effects and thus their incompatibility with the desired level of environmental and health protection. With the help of defined methods and criteria, it must be possible to clarify which reasons for concern can be considered as triggers for precautionary measures, what significance should be assigned to these 'reasons for concern' in each case, and what consequences should be drawn from them in precautionary risk management.

For the identification of reasons of concern, two perspectives are essential: on the one hand, the focus on the entity that affects the systems concerned, the agent, the

[10]In Germany, the Federal Institute for Risk Assessment (BfR) and the Central Commission for Biological Safety (ZKBS) should be mentioned, as should the Federal Office of Consumer Protection and Food Safety (BVL), the Federal Environment Agency (UBA) and the Federal Agency for Nature Conservation (BfN). At European level, the European Food Safety Authority (EFSA) and the European Environment Agency (EEA) deserve special mention. Internationally, the Conference of the Parties to the Convention on Biological Diversity with its Secretariat and its Subsidiary Body on Scientific, Technological and Technical Advice (SBSTTA) is of great importance.

technology, the intervention. This is done with the method of technology characterization using the criteria intensity and depth of intervention, resulting in technological power and range, and reliability. Then the view turns to the systems that may be affected. Their vulnerability and social criticality are investigated using the method of vulnerability analysis, with the focus on identifying highly vulnerable entities, weak points as well as tipping points as sources of surprise.

SPAGE technologies are currently still at a very early stage of innovation. Thus, the exact application objectives and application contexts or possibly affected systems are still largely unknown. Additionally, to the technology (the agent) these objectives and contexts have to be investigated as further sources of hazards, exposures and risks. The precaution-oriented prospective technology assessment in the phase of science and technology development therefore first concentrates on what can be already known, on the SPAGE technologies currently under development. The search for reasons for concern is therefore initially carried out within the framework of a technical characterisation and as soon as application objectives, contexts and systems are known to some extent, in the form of a vulnerability analysis of the potentially affected systems.

Dealing with Non-knowledge: Precautionary, Prospective Technology Assessment Versus Environmental Risk Assessment

Currently, the debate on health and environmental risk assessment of planned releases of genetically modified organisms (GMOs) focuses on the authorization of food processes and products, with EFSA and its 'Panel on Genetically Modified Organisms' as key actors. They act in a comparatively late stage of innovation. A precautionary prospective technology assessment must start much earlier, already in the research process and during the development of technologies.[11] This has the advantage that necessary changes of direction and switching to lower-risk development paths can be carried out comparatively easily as long as path dependencies have not been established by far-reaching investments. However, this also has the already mentioned disadvantage that hazards and exposures resulting from application goals cannot yet be adequately considered.

The methods for Environmental Risk Assessment (ERA)[12] of genetically modified organisms (GMOs), which was first described in the EU Directive 2001/18/EC and

---

[11]"Each phase of research activity - from developing a research plan to post-release surveillance - raises different levels of concern depending on the organism being modified and the type of gene drive being developed", determines the National Academies of Sciences (2016, p. 7).

[12]The National Academies of Sciences (2016) define the Environmental Risk Assessment as follows: "the study and use of probabilistic decision-making tools to evaluate the likely benefits and potential harms of a proposed activity on the well-being of humans and the environment, often under conditions of uncertainty" p. 105.

later in the EFSA documents (cf. in particular EU Directive 2001/18/EC, EFSA 2006, 2013a, b), unfortunately does not take into account the fundamentally different views of the agent on the one hand (technology characterisation) and the affected systems on the other (vulnerability analysis). Anyhow, in its methodological instructions ERA takes into account many of the important aspects from both points of view in a loose order. In their technology-related analytical steps, some of which are explicitly referred to as 'characterization', there are numerous overlaps with a precautionary, prospective technology characterization. However, there is a gap that cannot be easily bridged concerning the understanding of and dealing with uncertainties and lack of knowledge. Additionally, on the system-related side of the ERA, with regard to the analysis of affected systems and possible impacts in these, on the one hand points of reference can be identified and on the other hand further developments are necessary. Last but not least, a clear structure regarding the determination of vulnerabilities would approve the practiced ERA.

We distinguish two different types of vulnerability analysis an event-related impact oriented and a structure-related vulnerability analysis as explained in Chap. 4. The event-related vulnerability analysis (eVA) differentiates according to the

(a) Disturbing event/agent
(b) Exposure to the agent
(c) Sensitivity of the system to the agent
(d) Adaptive capacity of the system, its ability to process disturbance events (e.g. immune system).[13]

The vulnerability of the affected (eco)system to the agent (the SPAGES) then results from the integrated consideration of all four aspects. In fact, also the ERA pays great attention to exposure and sensitivity. The 'Guidance to develop specific protection goals options for environmental risk assessment at EFSA, in relation to biodiversity and ecosystem services' additionally addresses possible tipping points, the complexity and resilience of ecosystems, which are in the focus of the structure-related vulnerability analysis (EFSA 2006). However, this second approach, the structural vulnerability analysis (sVA), is neglected due to ERA's mainly event-related approach.[14] The sVA concentrates on weak points in the affected systems independently of possible disturbance events, which can be both punctual and creepingly continuous. It analyses which elements and relations of the system may draw back when it comes under pressure.[15] They all are about evidence for the presence of particularly sensitive or particularly critical system elements and relations or

[13]See von Gleich et al. (2010), Wachsmuth et al. (2012), Gößling-Reisemann et al. (2013).

[14]The eVA and sVA (cf. von Gleich et al. 2010; Wachsmuth et al. 2012; Gößling-Reisemann et al. 2013) cannot be elaborated further in the context of this text. However, important aspects of an event-related vulnerability analysis are contained in the reports on the case studies GMO olive fly and GMO rape seed (Chaps. 4 and 5), and important aspects of a structural vulnerability analysis are contained in the chapter on 'Tipping Points' (Chap. 2).

[15]Methodological role models for an sVA of socio-technical systems are engineering methods such as Failure Mode and Effect Analysis (FMEA) (cf. DIN e.V. 2006; Eberhard 2012), Fault Tree

the presence of particularly unstable or pre-tensed system states.[16] This subheading also includes the discussion about possible contamination of (particularly valuable?) ecosystems by GMOs. Can or must the transfer of modified genes to wild forms or the mere presence of GMOs in certain ecosystems already be regarded as a reason for concern?[17] In case of interventions into particularly sensitive, pre-tensed[18] or pre-damaged systems there are good reasons to expect that this will have far-reaching consequences. In such situations the implementation of the precautionary principle is recommended, and the corresponding indications of 'particularly alarming system conditions' are among the comprehensible and valid triggers (reasons for concern) for measures according to the precautionary principle. The sVA and the eVA are thus complementary and their approach is significantly more precise than that of the ERA due to the differentiated consideration of disturbance events, exposure, sensitivity and capability to adapt (eVA) as well as the analysis of internal weak points in the affected system (sVA). The technical characterization concentrates on the agent, the sVA concentrates on the affected systems and the eVA focuses on the interactions between the two.

The sVA plays a special role within the framework of a precautionary, prospective technology assessment and especially as a starting point for designing resilient systems. It is a method that opens options to minimize weak points, to increase the resilience of the systems concerned, to prepare them for possible surprises, for events and mechanisms of action, which are not yet known. This opens up a second way to practically reduce the realm of possibilities and thus the extent of ignorance regarding possible far-reaching disruptive events. The first strategy applies to the agent and focuses on substitution, on less depth of intervention, less powerful and far-reaching technologies, which reduce the range of possibilities of disruptive events. This was already elaborated in Chap. 7 and will be picked up again later on. The second strategy tries to reduce the realm of possibilities by strengthening the adaptive capacity and resilience of the systems concerned. Resilient systems are able to successfully cope with a wide range of even unknown disruptive events. This presupposes, however, that one is—not least also technically—in a position to relate precautionary risk management not only to the intervening technologies, but also to the affected

---

Analysis (FTA) (Böhnert 1992; Thums 2004) or regarding complex organisations e.g. bank stress tests (cf. e.g. Quagliariello 2009).

[16]Criticality refers to areas that are particularly important for social life or survival, e.g. nutrition, health, medicine. Sensitivity refers to particularly sensitive areas, phases or subgroups, e.g. pregnancy, previous damage or tipping points.

[17]The report of the National Academies of Sciences (2016) states: "The mere presence of the modified genetic element in other species could be considered an endpoint" p. 110 and further: "Because the goal of a gene-drive modified organism is to spread, and possibly persists, in the environment, the necessary ecological risk assessment is more similar to that used for invasive species, than for environmental assessment of genetically engineered organisms" p. 110, see also Landis (2004).

[18]In his presentation of the 'Adaptive Cycle' Holling gives an example of a pre-tensed system. A forest may have accumulated a great deal of energy in the form of wood during its conservation phase and is then in danger to suddenly burn down in case of ignition (release), cf. Holling (1986).

systems (their vulnerability, adaptive capacity or resilience).[19] Such a 'constructive precautionary approach', as Hansen et al. (2007, p. 400) call it, is totally out of reach of the strictly event-related ERA and the risk management based on it.

Conventional risk assessments focus on risks and not on precautions. According to the ERA-related understanding of risk, they aim at quantifiable statements in which risk is defined as the product of the amount of loss and the probability of occurrence. Risk analysis is then dependent on immense amounts of data. Though the deficiency of knowledge is omnipresent. But the latter is only understood as a deficiency that can be overcome by more research. There is no room for reflection on the fundamental limits of knowledge, e.g. the impossibility of predicting the effects of interventions in complex systems due to non-linearity, tipping points or emergence. There is certainly no room for reflections about ignorance, cluelessness and complete surprises,[20] about a non-knowledge of which we don't even know that we don't know (unknown unknowns Wynne 1992; Kerwin 2016).

In fact, the ERA also incorporates approaches that relate to minimizing non-knowledge. The ERA practitioners, especially at EFSA, are well aware that the possibilities for risk assessment based on predictions are clearly limited. In order to bridge the lack of knowledge, ERA uses a number of auxiliary constructions to get data after all (cf. e.g. EFSA 2010). Most important are the so-called 'concept of familiarity', the assessment of risks of GMOs by estimating the partially better known risks based on the donor organism, the recipient organism and the vector, and the 'comparative approach', the use of risk assessment with reference to already naturally occurring comparable organisms (non-GM-surrogates). The fact that ERA additionally makes use of a number of methodological steps of 'societal learning' shows that surprises are at least implicitly expected (cf. von Knies and Winter 2011, p. 8). To be mentioned here are the step-by-step procedure on the way towards release and the approach of post-market environmental monitoring (PMEM) as well as general surveillance (GS). All this, however, has the fatal disadvantage that learning may come too late and there may be no chance for corrective action.

---

[19] Stirling et al. (2006) speak of a "Resilience-focused (risk absorbing) Management Style" in this respect: "Improving capability to cope with surprises: diversity of means to accomplish desired benefits, avoiding high vulnerability, allowing for flexible responses, preparedness for adaptation" (p. 302). Strategies of an 'adaptive management' of socio-ecological systems, as they were developed after Holling (1978), Shea et al. (2002), Stankey et al. (2005), can also be considered. Specific references to design principles for increasing the resilience of socio-technical systems can be found in von Gleich and Giese (2019).

[20] Having in mind causes for surprises Stirling et al. (2006) mention "substantive novelty" or "unprecedented characteristics" (p. 294). Brooks (1986) developed a typology of surprises regarding the interaction between technology and society. He suggests three types of surprises: (1) unexpected discrete events, (2) discontinuities in long-term trends, and (3) emergence und sudden public awareness of new information. Filbee-Dexter et al. (2017) gave an interesting overview of 'ecological surprises' mostly combined with tipping points, phase transitions and feedback mechanisms (e.g. unanticipated behaviour or regime shift in ecological systems as collapse of coastal fisheries, state change during eutrophication of lakes, outbreaks of insects or species invasions). Close to the problem of total surprises are extremely improbable events, which Taleb (2007) called 'black swans'.

And all this deficiencies of the ERA approach are not adequately communicated at any point. This leaves the impression of an indefinitely optimistic quantitative 'risk assessment', which is aware only of the not-yet-known, which gets by completely without precautionary approaches and ignores essential references to possible surprises, to the remaining lack of knowledge, to unpredictability and ignorance (the unknown unknowns) (cf. Wynne 1992; Wehling 2009). If EFSAs guidance documents address uncertainties[21] and, to some extent, the limits of knowledge, then it is only in relation to the limits set on risk research by scarce resources (money and time). Statements about the limits of predictability in dealing with complex ecosystems—i.e. aspects of not being able to make predictions—are suppressed, as can be shown by two processes.[22] The 'Scientific Opinion on Guidance on the risk assessment of genetically modified microorganisms and their derived food and feed products' (n.d.) states: "Predicting impacts of GMMs and derived food or feed on complex ecosystems can be difficult due to continuous flux and spatial heterogeneities in ecosystems creating a myriad of potential microbial habitats in which interactions between GMMs and their products with the indigenous organisms and or abiotic components can take place. It is recognized that an environmental risk assessment cannot provide data of a GMM or their products which would cover all potential environmental habitats and conditions. Consideration of environmental impact (damage) should therefore focus on environments in which exposure is most likely or in which, when relevant, viable GMMs could potentially proliferate" (EFSA n. d., p. 38). These formulations are missing in the Guidance Document of 2006 on the same topic, which was finally published. Instead it reads: "Predicting impacts of GMMs and derived food or feed on complex ecosystems that are continually in flux is difficult and largely based on experiences with other introductions and an understanding of the robustness of ecosystems. It is recognized that an environmental risk assessment is limited by the nature, scale and location of experimental

[21] The most comprehensive and clearest summary is the "Generic list of common types of uncertainty affecting scientific assessments" EFSA (2018a, p. 19).

[22] The background of this attitude may be that lack of knowledge and ignorance are often put forward as reasons for the need to apply the precautionary principle. Whether the mere indication that complex systems are being interfered with and that not all possible reactions of these systems have yet been known and researched can suffice as a trigger for far-reaching precautionary measures is currently the subject of controversial debate. However, such an approach is already addressed in the Rio Declaration as a possibility: "In the case of measures relating to complex systems that have not yet been fully understood and for which the consequences of disruptions cannot yet be predicted, the precautionary approach could serve as a starting point" (Chapter 35, Paragraph 3 of Agenda 21). If one wants to justify the necessity of far-reaching precautionary measures, however, one should not stop to point out a lack of knowledge. Comprehensible knowledge about reasons for concern is necessary. Identifyable non-linearity, feedback mechanisms, bifurcations and tipping points, as well as pre-tensed and pre-damaged systems are reasons for concern recognizable through vulnerability analysis comprising the use of models, and they are also dependent on identifiable system architectures and system states. What is not comprehensible, however, is an argumentation in which the complexity paradigm is first relied upon to justify a lack of knowledge, and in which the argument suddenly jumps back into the reductionist paradigm and demands complete controllability. If one uses the complexity paradigm argumentatively, a demand for complete controllability does not make sense.

releases, which environments have been studied and the length of time the studies were conducted" (EFSA 2006, p. 59). In the preliminary version, the limits were still mentioned which, in principle, are set for predicting the consequences of interventions in complex systems. In the version finally adopted, on the other hand, EFSA returns to the position previously published several times, according to which the project is limited above all by the research expenditure.

This course of action is also repeated in the two-stage process of the development of the EFSA 'Guidance on the environmental risk assessment of genetically modified animals'. The precedent 'Scientific Opinion' mentions uncertainties due to assumptions and extrapolations, conflicting scientific literature and perspectives, and specified uncertainties, the latter divided into linguistic uncertainty (lack of linguistic precision), variability (in the subject area), and uncertainty caused by limitations of scientific knowledge and knowledge production such as motivational and systematic bias, censoring, measurement error, missing data, lack of suitable comparators or surrogates, and other causes of incomplete awareness, understanding and descriptions of a mechanism, process or system (i.e. model and scenario uncertainty)" (EFSA 2013a, p. 41f). Subsequently, methods for the reproducible identification and handling of these uncertainties are discussed. The knowledge problems arising from long-term exposures are addressed and reference is made to the auxiliary construction of the familiarity concept (EFSA 2013b, pp. 38f, 41, 163f). The following conclusion is then drawn: "ERA is often constrained/restricted by the available knowledge and experience of the GM animal and it can be difficult to predict and consider all potential future applications, production systems and receiving environments of the GM animal. Thus large-scale and long-term use of a GM animal could result in some effects which were not predictable at the time of the ERA or consent. Therefore, according to Directive 2001/18/EC, applicants are required to conduct general surveillance (GS) to detect unanticipated adverse effects on the environment" (EFSA 2013a, p. 44). Interestingly here too, another text appears in the later adopted and published version: "Overall, the results of the ERA will be subject to varying levels of uncertainty associated with factors such as (1) the availability of data and use of non-GM surrogates to inform the ERA, (2) the range of receiving environments in the EU where the GM animals are likely to be intentionally or accidentally released and (3) the diversity of management practices across EU regions. As far as possible, the overall conclusions of the ERA should specify under which conditions (e.g. receiving environments, management practices of the placing on the market, release and production) the risks/uncertainties identified are most likely to occur and clearly identify the factors/processes which might affect the conclusions of the ERA in order to make explicit the robustness of the conclusions of the ERA (EFSA 2013b, p. 30)". Here, too, the concession of limits of predictability is withdrawn in favour of a diversity that cannot be coped with in terms of effort and resources. Unknown unknowns are not mentioned. At least, an extreme extension of the spatio-temporal range of GMOs is addressed as an insight-limiting aspect, but as a way out, reference is again only made to ex-post observation via Post Market Environmental Monitoring and General Surveillance (EFSA 2013b, p. 163), which only takes effect when 'the child has, so to speak, already fallen into the well'. The biggest deficits of the ERA therefore exist in dealing with special forms of non-knowledge (non-determinacy, unknown unknowns and possible surprises).

This did not really change when EFSA started a comprehensive process to address uncertainty from 2014 onwards. In the two documents representing the preliminary outcome of this process possible surprises or unknown unknowns are not addressed, not even the difficulties of predicting the consequences of interventions in complex systems (see EFSA 2018a, b). The term complexity appears only once, but related to the complexity of the assessment process. Non-determinacy, non-linearity and tipping points do not occur as terms.[23] Unknown unknowns are mentioned, but EFSA simply excludes them from the 'Uncertainty Analysis'. "It is important to note that overall uncertainty cannot and does not include any information about unknown unknowns, i.e. uncertainties not known to the assessors. Since these are unknown, they cannot be either quantified or described" (EFSA 2018a, p. 34). EFSA acts according to the motto 'you can't say anything about non-knowledge, because it's non-knowledge'. Unknown unknowns may exist, but we're not in charge. It is not the risk assessors who are responsible for dealing with unknown unknowns, but only the risk managers.[24] "Decision-makers should understand that all assessments are conditional on the current state of scientific knowledge, and do not take account of 'unknown unknowns', and take this into account in decision-making (e.g. they might treat novel issues differently from those with a long history of scientific research)" (EFSAa, p. 35). Such statements do not, however, prevent EFSA from repeatedly stressing that risk assessors should identify all relevant sources of uncertainty and that the outcome of the risk assessment naturally includes statements about how problematic or harmless (no concern) certain possible developments are (EFSA 2018b, p. 34). Which, of course, must be related to the level of protection targeted in the EU (von Schomberg 2006, p. 25). A risk assessment is not possible without values. In addition, risk managers are also dependent on scientific preparatory work for a cost–benefit assessment.

[23]This contrasts with the 'Scoping paper' of the Group of Chief Scientific Advisors of the EU Commission. They distinguish between 'scientific uncertainty', 'indeterminacy' and 'ignorance' (2018, p. 3), which corresponds to our distinctions.

[24]"Deciding how much certainty is required or, equivalently, what level of uncertainty would warrant precautionary action, is the responsibility of decision-makers, not assessors" (EFSA 2018a, p. 16). Contrary to this, Sterling et al.: "It is clear that the precautionary principle is of relevance not only to the management, but also to the assessment of risk" (2006, p. 289). The EFSA statement also reveals a serious misunderstanding about the functioning of the precautionary principle. It is assumed that the level of uncertainty could be a trigger for precautionary measures. Rather, it is true that in essence the scale of a potential threat triggers precautionary measures, despite remaining uncertainties about the probability of occurrence and the precise outcome of the threat.

Depth of Intervention

Scientific uncertainty, not-yet-knowledge (in the sense of lacking research results), unpredictability (due to non-determinacy in complex systems) and ignorance (regarding complete surprises) are the previously mentioned forms of non-knowledge. In the Environmental Risk Assessment especially unpredictability and ignorance are not considered adequately or not at all. Now another important form of non-knowledge has to be mentioned: The technically generated non-knowledge.[25] It is generated by spatio-temporal delimited entities characterized by an expanded half-life, persistence, mobility, chain reactions or the ability of self-proliferation. The magnitude of this form of cluelessness about possible consequences and surprises comes about through the practical extension of the effectiveness and scope of the results of technological interventions, of their technical power and range in space and time. This enormously expands the realm of possibilities for unexpected events. Seen precisely, the technically generated expansion of the realm of ignorance is owed to the enabling of a not overseeable and incalculable magnitude of possible interactions of spatio-temporal delimited entities in countless system contexts.[26] Well known examples are unmanageable chain reactions in the field of nuclear and chemical technologies on the side of power and on the side of exposure extreme half-lives of substances in the environment (persistence or radioactivity) or GMOs (modified genetic elements), which are able to multiply and spread on their own. If a genetically modified organism is able to reproduce itself and is also mobile, there is a threat that it will tend to spread globally, not retrievably and irreversibly. Due to its expanded spatial and temporal range, this organism can emerge in an infinite number of diverse systems and contexts and enter into completely unpredictable and surprising interactions. The best known example of such a surprise from the realm of chemicals is the ozone-depleting effect of CFCs in the stratosphere, i.e. at an altitude of more than 10,000 m. The magnitude of ignorance about the possible consequences of such technological interventions will thus increase in proportion to their technical power and range. From this point of view, the fact that gene drives are explicitly produced for the purpose of rapidly spreading in populations must be classified as particularly concerning. Since this expansion of ignorance was created technically, however, it can also be reduced technically. An extreme spread can be technically reduced by switching to self-limiting reactions or to less invasive, less persistent, less mobile and reproductive GMOs. The focus of such a 'constructive precautionary approach' is

[25]This form of non-knowledge must not be confused with the omnipresent experience that scientific research on the one hand generates knowledge but on the other hand always raises new questions. This experience may be called 'science-based ignorance' (cf. Wehling 2009; Jäger and Scheringer 2009). Another form of 'producing ignorance' is already closer to this term. Ignorance is produced here not by technical means, but through political and communicative strategies with the aim of ignoring existing knowledge (cf. Aradau 2017; Proctor 1995). This form is the focus of so-called 'ignorance studies' and 'agnotology' (cf Gross and McGoey 2015; Proctor and Schiebinger 2008).

[26]This may include cause-effect relations not yet known (as in the case of CFCs) or interaction with tipping points and non-linearity in complex systems (as in the case of climate change) and even emergent behavior.

on substitution is on changing the character of the agent (see Chap. 7). This marks a clear difference from the widespread practice of reducing exposure to an unchanged agent by containment.[27] Anyhow, both of these strategies aim at reducing the range of possibilities.

The technical roots of the expansion of the realm of possibility and the associated magnitude of non-knowledge can be identified within the framework of technology characterization. This is done not least with the help of the technical evaluation criterion "depth of intervention". This criterion was introduced at the end of the 1980s (cf. von Gleich 1989), building on considerations by Günther Anders and Hans Jonas (cf. Anders 1958; Jonas 1979, 1985). It refers to technologies based on the mathematical-experimental natural sciences, for which a distinction between the level of phenomena and the level of laws of nature behind the phenomena is constitutive. Within this paradigm the laws of nature produce the natural phenomena and control them to a large extent. The atoms (or elementary particles) largely determine the physical properties of the physical objects, the molecular structures control the chemical properties of substances and also the genes are decisively involved in the control of the biological properties of organisms. Technologies that technically address (manipulate) such control structures as atoms/elementary particles, molecular structures or genes generate a significantly higher power over phenomena and more far-reaching consequences in comparison to techniques that only address (manipulate) directly perceptible phenomena, as has been the case for millennia with e.g. artisanal agricultural technology (e.g. in the form of breeding by selection). Breeding in which a greater variability is produced by irradiation or chemicals must therefore also be regarded as a deep intervention. A greater depth of intervention by addressing such 'control structures' can thus be identified as the basis for a broader technical power and range in space and time.[28] The basis of this approach to technology characterization is thus the insight that the magnitude of ignorance about possible consequences of deep interventions is not 'simply already there', but rather is generated and enormously expanded by the character (the depth) of the technical intervention.[29] The use of particularly powerful and far-reaching technologies increases the scope of what can happen and thus also the ignorance regarding possible consequences. Conversely, this range of possibilities and the combined extent of cluelessness regarding possible consequences can also be reduced by the use of techniques with a lower depth of

[27]This difference is also addressed as intrinsic biocontainment versus extrinsic and technical containment cf. European Science Foundation (ESF 2012).

[28]The criterion depth of intervention was successfully used within the framework of prospective technology assessment in areas such as synthetic chemistry (Böschen et al. 2003), nanotechnology (Rip 2006) and synthetic biology (Grunwald 2016). Related conceptualizations of the criterion depth of intervention in the biological field can be found in Deutscher Ethikrat (2011) as well as in Engelhardt et al. (2016).

[29]Wynne (2005) formulates this regarding the technical intervention at the gene level as follows: "The very idea of intervening in nature at the utterly novel genetic level, despite being championed both as a way of *increasing* knowledge, and as a new dawn of precision-biotechnology, introduces and releases onto society previously unencountered (and hitherto irrelevant) elements of the unknown, thus augmenting unpredictability and potential *lack* of control. Scientific research may not only diminish ignorance, but also thus amplify it too." p. 69.

intervention. If, as part of substitution efforts, we can select or develop agents that are degraded or die after a few hours or days, this will significantly reduce the extent of ignorance about possible consequences (see Chap. 7).

Identification of Substances of Very High Concern in REACH

Such considerations about technically generated non-knowledge are not new. They have already played an important role in the environmental policy debate on persistent industrial chemicals and in particular on stratospheric ozone depletion as a result of CFC release (cf. Scheringer 1996). And finally, based not least on the debate about persistent organic pollutants (POPs) in the oceans, they have also found their way into the European chemicals regulation under REACH, which classifies very persistent and bioaccumulative substances as 'substances of very high concern'.

Until then, the debate on possible triggers for precautionary measures had focused exclusively on environmental and health risks. Particularly widespread exposure to agents has not been accepted as a cause of concern and for precautionary measures so far. This changed with the REACH chemicals legislation. Article 1 of REACH states in point 3: "This Regulation is based on the principle that manufacturers, importers and downstream users must ensure that they manufacture, place on the market and use substances which do not adversely affect human health or the environment. Its provisions are based on the precautionary principle" (REACH Regulation 2006, p. 28).

The design of REACH then really endeavors to operationalize the precautionary principle and explicitly deals with the question of how precautionary measures are to be triggered. Point 69 of the list of recital grounds for the adoption of the Regulation states: "In order to ensure a sufficiently high level of protection of human health, including affected populations and, where appropriate, of certain vulnerable sub-groups, and of the environment, substances of very high concern should be treated with great care in accordance with the precautionary principle" (p. 14). The concept of 'substances of very high concern' must be emphasized in this formulation. According to REACH, 'substances of very high concern' are not only substances with a high hazard potential (in particular carcinogenic, mutagenic or toxic for reproduction), but also substances with an especially high exposure potential, namely particularly very persistent and bioaccumulative substances, even irrespective of any associated hazard hypothesis. This is the first important step that should be followed when implementing the precautionary principle in the governance and regulation of GMOs. The basis for this step towards a practically 'hazard-independent operationalization of the precautionary principle' is the acceptance that an enormous expansion of exposure expands the scope for unexpected interactions in the environment to the extreme and thus also the ignorance of possible surprises. Interestingly, this logical consideration can also be interpreted as a special form of the hazard hypothesis.[30]

Within REACH the characterization of a substance as being of very high concern has a precise and direct impact on risk management. This is the second important step that should be taken in the operationalization of the precautionary principle in the area of the GMO. The characterization of a substance as being 'of very high concern' leads to a general ban on its use (Art. 56) and to an explicit obligation to obtain authorization. The substance will be included in Annex XIV of the REACH Regulation. An authorization is only possible if it can be shown that the risk posed by this substance can be controlled throughout its entire life cycle. For substances for which no threshold value can be given,[31] exemptions are possible on the basis of a balance between benefit and risk potentials.[32] In addition, even after an exceptional authorization, still exists the minimization requirement, an obligation to search for less problematic alternatives (substitution), and an obligation to monitor the fate of the substance in the environment (Article 60 (10)). For the governance of chemicals, therefore, a comparatively easy way has been found to integrate the precautionary principle into the technology-related regulation following these two steps.

The particular practicability of this procedure lies in the fact that it first focuses on the properties of the agents. This has the advantage that one does not depend yet on precise toxicological knowledge and on the particularly complex vulnerability analysis of certain target systems. The procedure concentrates on the 'inherent properties' of the agents in the sense of their technical character. It is comparatively straightforward to determine the physico-chemical or biochemical properties of substances as being very persistent [CFCs have half-lives of up to 400 years (cf. Koch 1995, p. 259)] and as being very bioaccumulative. The extension of their range is in many cases a consequence of the depth of intervention during their production, the synthesis of non-natural, persistent and mobile chemicals.

However, also with REACH two problems have not yet been solved with regard to the implementation of the precautionary principle. REACH only starts late in the innovation process with the approval of chemicals and products. The constructive approach of precautionary risk management, the precautionary design of environmentally friendly chemicals,[33] is largely ignored. The quest for substitutes comes too

---

[30] Sterling therefore describes the exposure-oriented characteristics as "proxies for possible harm" Sterling (2016, p. 15).

[31] For persistence and bioaccumulation, no effect threshold can be given because these properties relate only to exposure. All the more a PEC/PNEC comparison is not possible due to the lack of a quantifiable impact threshold and the impossibility of determining an expected environmental concentration. On the other hand, emphasis is placed on the spatial and temporal decoupling of emission and possible effect (cf. Merenyi et al. 2011).

[32] For these substances "authorization can only be granted if it is demonstrated that the socio-economic benefits outweigh the risks to human health or the environment arising from the use of the substance and if no suitable alternative substances or technologies are available" (Art. 60, para. 4).

[33] Cf Hansen et al. (2007), e.g. with the goal 'benign by design', see Laber-Warren (2010), Leder et al. (2015).

late. And another necessary step is still missing. If applications are already known, an event-related eVA and a structural vulnerability analysis sVA should be established with a focus on sensitivity, adaptive capacity, criticality, non-linearity and tipping points. In this way, the classification of an agent as being 'of very high concern' as result of the technology characterization should be supplemented by the classification of (elements of) target system as 'being of very high concern' or being in a 'state of very high concern' from vulnerability analysis, each with corresponding consequences in risk management.

Operationalization of the Precautionary Principle in the Governance of GMOs Along the Lines of REACH

If the political will is there to do so, it should be comparatively easy to transfer the steps taken in REACH to operationalize the precautionary principle into the governance and regulation of genetic engineering. The hazard and exposure assessment would initially focus on the characterization of GMOs and genetic engineering constructs. The intensity and depth of intervention and its two consequences, technological power and range, initially will play a central role. As a result of this step alone, GMOs or genetic engineering constructs can be characterized as being of high concern. However, it has to be taken into account that a technology characterization with regard to the release of living organisms is associated with additional challenges in comparison to chemical substances. It is true that even chemicals must be expected to change after release (ageing, oxidation, metabolisation). However, organisms have a significantly higher ontogenetic and phylogenetic plasticity. Nevertheless, the technical characterization of the GMO is a rather simple and low-cost procedure in comparison to the vulnerability analysis of the affected systems (and also in comparison to the steps required by the ERA). And the determination of the persistence and invasiveness of SPAGEs and GMO, their ability to reproduce themselves, to survive (evolutionary fitness) as well as their ability to spread over time and space is already a subject of the ERA.

The Regulation (EC) No. 1829/2003 of the European Parliament and of the Council of 22 September 2003 on genetically modified food and feed and the guidance documents of the EFSA describe in detail which risk-relevant analyses an applicant has to carry out and on which questions he should provide information (Regulation (EC) No. 1829/2003). The Environmental Risk Assessment related to the release of GMOs into the environment also begins with certain forms of technical characterization (characteristics of GMOs and releases).[34] The extension of the technology

[34]In particular, the following aspects are considered: the recipient or parental organism(s); the genetic modification(s), be it inclusion or deletion of genetic material; and relevant information on the vector and the donor; the GMO (including phenotypic and genetic instability); the intended release or use including its scale, cf. Regulation (EC) No. 1829/2003.

characterization by the criteria depth and intensity of intervention (regarding intensity of intervention see Chap. 1) and their consequences, technological power and range and additional liability should therefore be feasible by comparatively simple means. Aspects of intervention intensity are already represented by the 'scale of the intended release'. Almost all essential points of a (precaution-oriented) hazard and exposure analysis are already addressed, however with quite different intensity and, if Annex II of Regulation No. 1829/2003 is consulted, with a very different degree of detail. Very detailed instructions and scientific methods exist for technology characterization (partly referred to as molecular characterization) and for toxicological analyses (with standards also for allergy and nutrition physiology analyses). This also applies in part to comparative analyses with a "conventional counterpart". On the other hand, the area of environmental impacts is extremely under-represented, which is probably mainly due to the focus on food and feed. There are few indications of ecosystem effects and no indications of possible influences on biodiversity. However, these are discussed in detail in EFSA 2016. Comparatively great attention is paid to the reliability of genetic engineering methods and their undesirable side effects and consequences. What is striking, however, is that their spatial and temporal significance is not queried. There is a lack of specifications how to accurately identify and assess a "potential risk associated with horizontal gene transfer". The technical range and exposure is also given attention in the form of an exposure characterization, not only with regard to the quantitative or estimable aspects "predicted consumption, probable individual and age-specific intake", "recommendations for use, handling", but also with regard to aspects of technology characterization. "the spread of the GMO(s) in the environment (persistence and invasiveness, biological fitness, pathways of dispersal, reproductive, survival and dormant forms); interactions with target or non-target organisms; vertical or horizontal gene transfer; exposure to humans to animals; competition for natural resources like soil, area, water, light, displacement of natural populations of other organisms; delivery of toxic substances; different growth patterns)" cf. Regulation (EC) No. 1829/2003; EFSA 2006, 2013, but again no mention is made of the spatial and temporal implications, possible global spread and irreversibility of releases. The environmental risk assessment shall examine "possible changes in the interactions between the GM plant and its biotic environment resulting from the genetic modification, persistence and invasiveness, selection advantage or disadvantage, gene transfer potential, interactions between the GM plant and target organisms", "interactions between the GM plant and non-target organisms“. However, even here is a lack of guidance on how this requested information can actually be obtained.

In order not to be misunderstood, these enumerations serve above all to show that many aspects are already taken into account which are also important for a classification of 'genetic engineering constructs' or 'GMOs ' as being of 'very high concern'. In contrast to widespread criticism of the ERA, the approach taken here to implement the precautionary principle does not essentially aim at 'more knowledge' about what is not yet known or adequately considered. Rather, it aims at a different weighting of the already existing and comparatively easily accessible findings and, above all, at clear precautionary consequences from the latter.[35]

Even if one is unwilling to follow the criterion 'depth of intervention', the following central criteria remain for the identification of constructs and GMOs as being of high concern:

- Enormous **technological power** (e.g. virulence) combined with insufficient technical maturity and reliability.[36]
- Ability for **self-propagation** whereby a distinction must be made here between the propagation rate and the generation times of GDO as well as the ability to overcome Mendelian inheritance rules in GDO
- Genetic **Fitness** of the population (cf. Barker 2009)
- **Invasiveness** or threshold value of propagation, colonization
- **Persistence**, capability to persist and spread over time
- **Mobility**, capability of spatial propagation
- Potential for **vertical gene transfer**/hybridization potential e.g. use of conserved sequences as target loci for integration of homing endonuclease based gene drives
- Potential for **horizontal gene transfer**.

Conclusion

Important points of a precautionary technology characterization and vulnerability analysis are already addressed in the current guidelines and regulations. However, this should be better structured and operationalized. The main task of integrating the precautionary principle into the governance of new genetic engineering, however, is not to collect countless additional data. The identification and classification of GMO or genetic engineering constructs as being of high concern on the one hand and the improvement of the interface between hazard and exposure evaluation and risk management on the other hand by drawing clear precautionary consequences from this classification are of crucial importance. In the current situation, it is completely unclear what follows from the statements required by the applicant on 'risk characterization', 'exposure assessment' or 'environmental compatibility'.

---

[35]For the same reason, the debate as to whether relying on quantification is problematic and whether more qualitative information needs to be taken into account in the ERA does not play a major role here either. Rather, there is agreement that the evidence on the basis of which precautionary measures are to be triggered should be scientifically comprehensible (and, if possible, quantifiable). However, non-quantifiability should not lead to the exclusion of comprehensible indicators. Finally, for the same reason, the important role of participation in the implementation of the precautionary principle is not discussed here.

[36]Also with regard to technical reliability, there are quite a few reference points in the specifications of the ERA (e.g. genetic stability of the insert, stability and expression of the transformation events, biological plasticity...).

Five Steps Towards Integrating the Precautionary Principle into the Governance of SPAGE

1. **Pursue a constructive precautionary approach**

Precautionary risk assessment and precautionary risk management should not begin with product approval. They must begin already in the phase of research and development. Early in the innovation process, when path dependencies have not yet been consolidated by far-reaching investments, corrections, substitutions and the development of lower-risk development paths are much easier. However, due to the much greater degree of lack of knowledge about impacts at this stage, we should not talk about risks (whose assessment at this stage would require information that cannot be obtained), but about risk potentials and the underlying hazard and exposure potentials. Options for measures to influence research and development lie in precautionary risk research, in target-oriented funding programs for lower-risk alternatives,[37] in competitions and prizes, but also in the transparency of processes and opportunities for participation. The targeted promotion of a low-risk design of genetic and biotechnological constructs (benign by design) is particularly important. If it does not want to always come too late and intervene restrictively, the orientation to such a design is an indispensable approach of the precaution-oriented risk management.

2. **Consider all kinds of non-knowledge**

Precautionary measures are dependent on the generation of precautionary knowledge. Knowledge related to precaution should be able to understand the extent and possible consequences of a lack of knowledge in the form of comprehensible reasons for concern. All forms of non-knowledge must be taken into account, not only the uncertainties currently mentioned in the ERA and in approval procedures, but also unpredictability (the limits of prediction in dealing with complex systems due to non-determinacy), complete ignorance of possible surprises (unknown unknowns), and last but not least the technically produced extension of the realm of possibility and the thus extended ignorance.

3. **Pursue technology characterization and vulnerability analysis**

Technology characterization can start particularly early in the innovation process, even in the phase in which gene drives are currently being developed, in which hardly any applications are yet on the market. With the criteria of intensity and depth

[37] Wynne (2005) points out that the UK Biotechnology and Biological Sciences Research Council recommends "genomically-informed but non-transgenic approaches to crop science research", as a kind of research that is more likely to meet society's expectations in its rejection of green genetic engineering (BBSRC 2004, p. 35). In its report on 'Genomics and Crop Plant Science in Europe', the European Academies Science Advisory Council also recommended "non-reductionist functional genomics informing marker-assisted selection for identifying non-GM, naturally occurring desired crop traits" (EASAC 2004, p. 7). Wynne stated "A previously invisible alternative scientific trajectory marginalized by the exclusive GM paradigm, came rapidly to the fore, as the necessity suddenly arose" (p. 79f).

of intervention as well as liability, the focus is not first on effects, but on the character of the intervention, which produces these effects in the first place, especially technological power and range. Particularly high technological power (up to the triggering of chain reactions) and particularly high range of exposure (up to globality and irreversibility) through GMO and genetic engineering constructs can be characterized as being 'of high concern'. The criteria intensity and depth of intervention with the dimensions of technological power and range meet the requirements for the performance of precautionary criteria and reasons for concern, with regard to the degree of seriousness of hazard and exposure potentials, the magnitude of the possible consequences and the extension of the ignorance generated by the depth of intervention. In addition, it provides indications as to the direction in which lower-risk alternatives can be successfully sought.

In structural vulnerability analysis, it is important to identify particularly critical and sensitive systems or system elements, to identify tipping points and threatening phase transitions and bifurcations. If intervention is planned in systems which are critical for society (e.g. nutrition, health), which are pre-loaded or pre-tensed or include tipping points, then intervention in these systems can be characterized as being of high concern.

Further work is needed on the two methodological approaches of technology characterization and vulnerability analysis, as well as on the criteria depth and intensity of intervention, and on further indications of serious hazard and exposure potentials.

4. **Exposure is just as important as hazard**

Exposure after releases into the environment must be given as much attention as to the hazard dimension. This has now become established in the risk governance of chemicals. Very persistent and very bioaccumulative chemicals are classified under REACH as substances of very high concern. Thus an extreme exposure potential, even without an associated hazard hypothesis, is considered to be of very high concern. Regarding an independent spread of GMOs or their genes in ecosystems, the term 'so what?' is still all too often used (von Schomberg 2006, p. 24). The ability of GMOs and genetic engineering constructs to spread in the environment must therefore be anchored as a major concern in the regulation of genetic engineering. The underlying risk hypothesis refers to the fact that the temporal and spatial extension of the presence of such persistent and invasive constructs enormously increases the likelihood of their interaction with different elements and relations in different ecosystem contexts and can thus lead to major surprises, as had to be learned from the example of CFCs. Minimizing exposures is therefore a promising approach to dealing with unknown unknowns.

5. **Improving the link between hazard and exposure evaluation and risk management**

The interface between hazard and exposure assessment and evaluation and risk management needs to be improved. The characterization of a GMO or construct as being of 'very high concern' should lead to the same consequences as under REACH, i.e.

a ban on use, an authorization requirement with exemptions and an active search for lower-risk alternatives.

References

Ammann, D., Hilbeck, A., Lanzrein, B., Hübner, P., & Oehen, B. (2007). Procedure for the implementation of the precautionary principle in biosafety commissions. *Journal of Risk Research, 10*(4), 487–501.

Anders, G. (1958). Die Antiquiertheit des Menschen. *Über die Seele im Zeitalter der zweiten industriellen Revolution*. München.

Barker, J. S. F. (2009). Defining fitness in natural and domesticated populations. In: J. van der Werf (Ed.), *Adaptation and fitness in animal populations* (pp. 3–14). Heidelberg: Springer-Verlag.

Böhnert, R. (1992). *Bauteil-und Anlagensicherheit*. Würzburg: Vogel.

Böschen, S., Lenoir, D., & Scheringer, M. (2003). Sustainable chemistry: Starting point and prospects. *Naturwissenschaften, 90*(3), 93–103.

Caplan, A. L., Parent, B., Shen, M., & Plunkett, C. (2015). No time to waste—The ethical challenges created by CRISPR. *EMBO Reports, 16*(11), 1421–1426.

Commission of the European Communities. (2000).*Communication from the commission on the precautionary principle*, Brussels. https://eur-lex.europa.eu/legal-content/EN/TXT/PDF/?uri=COM:2000:0001:FIN.

Core Working Group on Guidance for Contained Field Trials (multi-authored including Benedict, M.Q). (2008). Guidance for contained field trials of vector mosquitoes engineered to contain a gene drive system: Recommendations of a scientific working group. *Vector-Borne and Zoonotic Diseases, 8*(2), 127–166.

Deutscher Ethikrat (German Ethics Council). (2011). *Human–animal mixtures in research—Opinion*. Berlin.

DIN e.V. (2006). *Analysetechniken für die Funktionsfähigkeit von Systemen – Verfahren für die Fehlzustandsart- und -auswirkungsanalyse (FMEA)*. Berlin: European Union: Deutsches Institut für Normung e.V.

Eberhard, O. (2012). *Risikobeurteilung mit FMEA, expert Verlag Tübingen*.

EC DIRECTIVE 2001/18/EC of the European Parliament and of the Council of 12 March 2001 on the deliberate release into the environment of genetically modified organisms and repealing Council Directive 90/220/EEC. https://www.epa.ie/pubs/legislation/geneticallymodifiedorganismsgmo/2001-18%20Directive_consolidated.pdf.

Eckerstorfer, M., Heissenberger, A., & Gaugitsch, H. (2010). *Considerations for a precautionary approach in GMO policy*, Umweltbundesamt Report 0233, Vienna.

European Food Safety Authority EFSA (o. J.) EFSA Panel on Genetically Modified Organisms (GMO); Scientific Opinion on Guidance on the risk assessment of genetically modified microorganisms and their derived food and feed products. *EFSA Journal 20YY, volume*(issue), NNNN. [52 pp.]. https://doi.org/10.2903/j.efsa.20YY.NNNN. Available online https://www.efsa.europa.eu/efsajournal.htm.

EFSA. (2006). EFSA Guidance document of the scientific panel on genetically modified organisms for the risk assessment of genetically modified microorganisms and their derived products intended for food and feed use. *The EFSA Journal, 2006*(374), 1–115.

EFSA. (2010). Guidance on the environmental risk assessment of genetically modified plants. *EFSA Journal, 8*(11):1879

EFSA. (2013a). SCIENTIFIC OPINION Guidance on the environmental risk assessment of genetically modified animals. *EFSA Journal 20YY, volume* (issue), NNNN.

EFSA. (2013b). SCIENTIFIC OPINION Guidance on the environmental risk assessment of genetically modified animals. *EFSA Journal 2013, 11*(5), 3200, S. 45.

EFSA. (2016). *Guidance to develop specific protection goals options for environmental risk assessment at EFSA, in relation to biodiversity and ecosystem services*. https://doi.org/10.2903/j.efsa.2016.4499.

EFSA Scientific Committee, Benford, D., Halldorsson, T., Jeger, M. J., Knutsen, H. K., & More, S., et al. (2018a). Principles and methods behind EFSA's Guidance on uncertainty analysis in scientific assessment. Scientific opinion. *EFSA Journal 2018, 16*(1), 5122, 282. https://doi.org/10.2903/j.efsa.2018.5122.

EFSA Scientific Committee, Benford, D., Halldorsson, T., Jeger, M. J., Knutsen, H. K., & More, S., et al. (2018b). Guidance on uncertainty analysis in scientific assessments. *EFSA Journal 2018, 16*(1), 5123, 39. https://doi.org/10.2903/j.efsa.2018.5123.

Engelhard, M., Bölker, M., Budica, N. (2016). Old and new risks in synthetic biology: Topics and tools for discussion. In *Synthetic biology analysed*. Tools for Discussion and Evaluation (pp. 51–69). Cham: Springer.

European Commission. (2017). Science for environment policy. The precautionary principle: Decision making under uncertainty. Future Brief 18. Produced for the European Commission DG Environment by the Science Communication Unit, UWE, Bristol. http://ec.europa.eu/environment/integration/research/newsalert/pdf/precautionary_principle_decision_making_under_uncertainty_FB18_en.pdf.

European Commission. (2000). Mitteilung der Kommission vom 2. Februar 2000 zur Anwendbarkeit des Vorsorgeprinzips, Brüssel. https://eur-lex.europa.eu/legal-content/DE/TXT/PDF/?uri=CELEX:52000DC0001&from=DE.

European Environment Agency. (2002). Late lessons from early warnings: The precautionary principle 1896–2000. Environmental Issue Report, 22.

European Science Foundation (EFS). (2012). *ESF/LESC Strategic Workshop on biological containment of synthetic microorganisms: Science and policy*, Heidelberg (Germany), 13–14 November 2012. https://www.embo.org/documents/science_policy/biocontainment_ESF_EMBO_2012_workshop_report.pdf.

Fisher, E., Jones, J., & von Schomberg, R. (Eds.). (2006). *Implementing the precautionary principle—Perspectives and prospects*. Cheltenham: Edward Elgar.

Gößling-Reisemann, S., von Gleich, A., Stührmann, S., & Wachsmuth, J. (2013). Climate change and structural vulnerability of a metropolitan energy supply system—The case of Bremen-Oldenburg in Northwest Germany. *Journal of Industrial Ecology, 17*(6), 846–858.

Gross, M., & McGoey, L. (Eds.). (2015). *Routledge international handbook of ignorance studies*. London: Routledge.

Group of Chief Scientific Advisors. (2018). *EU Commission Scientific Advice Mechanism—Scoping paper—Making sense of science under conditions of complexity and uncertainty*, 1 February 2018. https://ec.europa.eu/research/sam/pdf/meetings/hlg_sam_scoping_paper_science.pdf#view=fit&pagemode=none.

Grunwald, A., (2016). Synthetic biology: Seeking for orientation in the absence of valid prospective knowledge and of common values. In S. O. Hansson, G. Hirsch-Hadorn (Eds.), *The argumentative turn in policy analysis*. Cham: Springer.

Hansen, S. F., Carlsen, L., & Tickner, J. A. (2007). Chemicals regulation and precaution: Does REACH really incorporate the precautionary principle. *Environmental Science & Policy, 10*, 359–404

Holling, C. S. (1978). *Adaptive environmental assessment and management*. Wiley.

Holling, C. S. (1986). The resilience of terrestrial ecosystems: Local surprise and global change. In W. C. Clark & R. E. Munn (Eds.), *Sustainable development of the biosphere: Interactions between the world economy and the global environment* (pp. 292–317). Cambridge, UK: Cambridge University Press.

Ishii, T., & Araki, M. (2016). Consumer acceptance of food crops developed by genome editing. *Plant Cell Reports, 35*(7):1507–1518.

Jäger, J., & Scheringer, M. (2009). Von Begriffsbestimmungen des Nichtwissens zur Umsetzung des Vorsorgeprinzips. *Erwägen Wissen Ethik, 20*(1), 129–132.

Jonas, H. (1979). *Das Prinzip Verantwortung*. Frankfurt/M: Versuch einer Ethik für die technologische Zivilisation.

Jonas, H. (1985). *Technik, Medizin und Ethik*. Frankfurt/M.

Kerwin, A. (2016). None too solid. *Knowledge, 15*(2):166–185

Koch, R. (1995). *Umweltchemikalien – Physikalisch-chemische Daten, Toxizitäten, Grenz- und Richtwerte, Umweltverhalten*. Weinheim: VCH.

Laber-Warren, E. (2010). *Green chemistry: Scientists devise new 'benign by design' drugs, paints, pesticides and more, new scientist*, May 2010.

Landis, W. G. (2004). Ecological risk assessment conceptual model formulation for nonindigenous species. *Risk Analysis, 24*(4), 847–858.

Leder, C., Rastogi, T.; Kümmerer, K. (2015). Putting benign by design into practice-novel concepts for green and sustainable pharmacy: Designing green drug derivatives by non-targeted synthesis and screening for biodegradability. *Sustainable Chemistry and Pharmacy, 2*, 31–36.

Merenyi, S., Kleihauer, S., Führ, M., Hermann, A., Bunke, D., & Reihlen, A., et al. (2011). *"Wirksame Kontrolle" von besonders besorgniserregenden Stoffen (SVHC) ohne Wirkschwelle im Rahmen der Zulassung nach REACH*, Abschlussbericht des Projekts, Umweltforschungsplan – FKZ 206 67 460/02.

National Academies of Sciences, Engineering, and Medicine. (2016). *Gene drives on the horizon. Advancing science, navigating uncertainty, and aligning research with public values*. Washington, DC: The National Academies Press.

Oye, K. A., Esvelt, K., Appleton, E., Catteruccia, F., Church, G., & Kuiken T., et al. (2014). Biotechnology. Regulating gene drives. *Science,345*(6197), 626–628.

Persson, E. (2017). What are the core ideas behind the precautionary principle?. *Science of The Total Environment* 557–558:134–141.

Proctor, R. N. (1995). *Cancer wars*. New York: Basic Books.

Proctor, R. N., & Schiebinger L. (2008). *Agnotology—The making and unmaking of ignorance*. Stanford University Press

Quagliariello, M. (Ed.). (2009). *Stress-testing the banking system: Methodologies and applications*. Cambridge (UK): Cambridge University Press.

REACH Verordnung. (2006). Verordnung (EG) Nr. 1907/2006 Registration, Evaluation, Authorisation and Restriction of Chemicals. https://eur-lex.europa.eu/LexUriServ/LexUriServ.do?uri=CONSLEG:2006R1907:20121009:DE:PDF.

Renn, O., Dreyer, M., Klinke, A., Losert, C., Stirling, A., & van Zwanenberg, P., et al. (2003). *The application of the precautionary principle in the European Union: Regulatory strategies and research needs to compose and specify a European policy on the application of the precautionary principle (PrecauPri). Stuttgart: Centre for Technology Assessment*. https://www.sussex.ac.uk/spru/environment/precaupripdfs.html.

Rip, A. (2006). The tension between fiction and precaution in nanotechnology. In: E. C. Fisher, I. S. Jones, & R. von Schomberg (Eds.), *Implementing the precautionary principle: Perspectives and prospects* (p. 278). Cheltenham: Edward Elgar.

Roller, G. (2005). Die Genehmigung zum Inverkehrbringen gentechnisch veränderter Produkte und ihre Anpassung an Änderungen des Standes der Wissenschaft. *Zeitschrift für Umweltrecht, 3/2005*, 113–119.

Scheringer, M. (1996). Persistence and spatial range as endpoints of an exposure-based assessment of organic chemicals. *Environmental Science and Technology, 30*(5), 1652–1659. https://doi.org/10.1021/es9506418.

Shea, K., Possingham, H. P., Murdoch, W. W., & Roush, R. (2002). Active adaptive management in insect pest and weed control: Intervention with a plan for learning. *Ecological Applications., 12*(3), 927–936. https://doi.org/10.1890/1051-0761(2002)012[0927:AAMIIP]2.0.CO;2.

Simon, S., Otto, M., & Engelhard, M. (2018). Synthetic gene drive: Between continuity and novelty—Crucial differences between gene drive and genetically modified organisms require an adapted risk assessment for their use. *EMBO Reports, 19*, e45760.

Stankey, G. H., Clark, R. N., & Bormann, B. T. (2005). Adaptive management of natural resources: Theory, concepts, and management institutions. Gen. Tech. Rep. Pnw-Gtr-654 (pp. 73, 654–673). Portland, OR: U.S. Department of Agriculture, Forest Service, Pacific Northwest Research Station. https://doi.org/10.2737/PNW-GTR-654

Stirling, A., Renn, O., & van Zwanenberg, P. (2006). A framework for the precautionary governance of food safety: Integrating science and participation in the social appraisal of risk.

Stirling, A. (2016). *Precaution in the Governance of Technology*. SPRU working paper series SWPS 2016-14 University of Sussex. http://sro.sussex.ac.uk/id/eprint/69089/1/2016_14_SWPS-Stirling.pdf.

Taleb, N. N. (2007). The Black Swan: The Impact of the Highly Improbable. New York: Random House

Thums, A. (2004). *Formale Fehlerbaumanalyse*. University of Augsburg.

TFEU. (2007). *Treaty of the Functioning of the European Union*. https://eur-lex.europa.eu/legal-content/EN/TXT/PDF/?uri=CELEX:12012E/TXT&from=EN.

UNEP. (2000). *Cartagena protocol on biosafety to the convention on biological diversity*. https://bch.cbd.int/protocol/text/CartagenaProtocolonBiosafety.

UNCED. (1992). *Rio Declaration on Environment and Development*. https://www.unesco.org/education/pdf/RIO_E.PDF.

Verordnung (EG) Nr. 1829/2003 des Europäischen Parlaments und des Rates vom 22. September 2003 über genetisch veränderte Lebensmittel und Futtermittel. https://eur-lex.europa.eu/legal-content/DE/ALL/?uri=celex:32003R1829.

Verordnung (EG) Nr. 1946/2003 des Europäischen Parlaments und des Rates vom 15. Juli 2003 über grenzüberschreitende Verbringungen genetisch veränderter Organismen Verordnung. https://eur-lex.europa.eu/legal-content/DE/ALL/?uri=CELEX%3A32003R1946.

von Gleich, A. (1989). Der wissenschaftliche Umgang mit der Natur - Über die Vielfalt harter und sanfter Naturwissenschaften Campus Verlag Frankfurt/New York.

von Gleich, A., Gößling-Reisemann, S., Stührmann, S., & Woizescke, P. (2010). Resilienz als Leitkonzept – Vulnerabilität als analytische Kategorie. In K. Fichter, A. von Gleich, R. Pfriem, & B. Siebenhüner (Eds.), *Theoretische Grundlagen für erfolgreiche Klimaanpassungsstrategien*. nordwest2050-Berichte 1, Bremen/Oldenburg.

von Gleich, A., Giese, B. (2019). Resilient systems as a biomimetic guiding concept. In M. Ruth, S. Gößling-Reisemann (Eds.), *Handbook on resilience of socio-technical systems*. Cheltenham, Northampton: Edward Elgar Publ.

von Knies, C., Winter, G. (2011). *The structuring of GMO release and evaluation in EU law. Biotechnology Journal 2011*, 6.

von Schomberg, R. (2006). The precautionary principle and its normative challenges. In E. Fisher et al. (Eds.), *Implementing the precautionary principle—Perspectives and prospects*. Cheltenham: Edward Elgar.

Wachsmuth, J., von Gleich, A., Gößling-Reisemann, S., Lutz-Kunisch, B., & Stührmann, S. (2012). Sektorale Vulnerabilität: Energiewirtschaft. In B. Schuchardt, S. Wittig (Hrsg.), *Vulnerabilität der Metropolregion Bremen-Oldenburg gegenüber dem Klimawandel (Synthesebericht)*. nordwest2050-Berichte Heft 2, Projektkonsortium 'nordwest2050' (pp. 95–112). Bremen/Oldenburg.

Wehling, P. (2009). *Nichtwissen – Bestimmungen, Abgrenzungen, Bewertungen*. In *Erwägen Wissen Ethik* 2009, Jg. 20 Ausgabe 1, S. 95–106.

Winter, G. (2016). In search for a legal framework for synthetic biology. In M. Engelhard (Ed.), *Synthetic biology analysed—Tools for discussion and evaluation*. Switzerland: Springer.

Wynne, B. (1992). Uncertainty and environmental learning. Reconceiving science and policy in the preventive paradigm. *Global Environmental Change, 2*(2), 111–127.

Wynne, B. (2005). Reflexing complexity—Post-genomic knowledge and reductionist returns in public science. *Theory, Culture & Society, 22*(5), 67–94.

Summary

Arnim von Gleich

With the new genetic techniques and especially with the targeted release of organisms in which self-replicating artificial genetic elements (SPAGE) are implemented, a qualitatively new stage in technology development has been reached. This applies both with regard to their technical efficacy and to their spread and exposure in the environment. Their enhanced technical qualities have consequences both in terms of risk management and risk governance and in terms of ethics and social acceptance. In particular, within the group of SPAGE there are quite different technologies. The GeneTip project focused on technologies with a dominant lethal gene (RIDL), meiotic drives (MD e.g. certain X-shredders), killer rescue technology, MEDEA (Maternal-Effect Dominant Embryonic Arrest), underdominance (UD) and endonuclease genes (HEG) with CRISPR/Cas9 systems.

Most SPAGEs—including gene drives in particular—are still at the beginning of their development process. The proof of concept still plays a central role. It is therefore often more a matter of science impact assessment than technology impact assessment. Starting the assessment at an early stage of innovation has to struggle with enormous knowledge problems. On the other hand, the scope for corrective measures is particularly large as long as path dependencies have not yet been formed. Early in the innovation process, a prospective technology assessment is necessary that is able to deal adequately with the enormous amount of non-knowledge about possible consequences. It therefore focuses on what is already known so early in the innovation process, the technology itself.

In the GeneTip project, a comparative prospective assessment of the hazard and exposure potentialsof various SPAGE technologies was carried out. For this purpose, the method of technology characterisation was applied, taking into account the criteria depth of intervention (divided into hazard and exposure potentials), intensity of intervention (divided into release quantity and number of releases), reliability (failures and side effects) as well as corrigibility (in case something goes wrong). The hazard potentials are essentially dependent on the application targets and the genetic information of the gene drives specially designed for these targets. For this reason, individual case studies such as those we have conducted for the olive fly and rapeseed are required for the investigation of the hazard potentials. With regard to the exposure potentials, however, more far-reaching generalizations were possible with regard to the various SPAGE technologies, supported by corresponding modelling approaches. Here, clear differences between the different SPAGE technologies could be observed, not only with regard to exposure, but also with regard to corrigibility and retrievability, cf. Table 2 in the Technology Characterisation chapter and Frieß et al. (2019).

These results can be seen as a starting point for practical technology choice and for the development of a more precautionary technology design. In the GeneTip project, the results of the technology characterisation were therefore also used as starting points for the identification of lower-risktechnology development pathways with which comparable technical goals can be achieved.

The hazard and exposure analysis and assessment of technologies must be twofold. On the one hand, it has to focus on the already mentioned side of the technology, which has been examined with the method of technology characterization. On the other hand—where application perspectives are already foreseeable—it has to focus on the side of the possibly affected systems. These were investigated using the method of vulnerability analysis method. Within the framework of vulnerability analysis, two approaches were distinguished: an event-related approach based on the motto 'what if?' and a structural approach based on the motto 'whatever may come'. The latter, independent of possible disturbance events, searches for weak points in the affected system at which the system will surrender if it comes under pressure. In this context, the search was also made for critical system elements, which are indispensable for important system services, and for tipping points, which can lead to far-reaching phase transitions and non-linearities. Tipping points are critical system states in which the smallest impulses or gradient shifts can have far-reaching consequences. As precautionary measures with regard to possible tipping points, (a) a systematization of their manifestations could be identified (seesaws, outbreaks, domino effects, excitable media, percolation, threshold effect, phase transitions, hysteresis, bifurcation, etc.), (b) tipping elements and tipping mechanisms could be identified within the affected systems on different hierarchical levels (molecular, organismic, population, biom, evolution), (c) if possible, early warningsignals could be identified that can indicate an approximation of system dynamics to tipping points (sluggish reactions, increased autocorrelation, etc.) and (c) approaches to tipping points could be minimized or avoided by precautionary risk management (comparable to the 2 °C target in climate precaution) or a precautionary more resilient design of the affected systems.

All this enables a first step towards precautionary preparation not only for known and expected disruptive events, but also for possible surprises, i.e. for disruptive events that are not expected or cannot be expected. This opens up perspectives far beyond classical risk management, which usually follows the 'what-if' approach. Extended precaution-oriented risk management also prepares for the unexpected and surprising, on the one hand for extremely unlikely events that are treated as negligible in classical risk management (the so-called black swans by Taleb 2007) and on the other hand for real surprises, i.e. the so-called unknown unknowns.

In two case studies, both the potential benefits and the risks of possible applications of gene drives in olive flies and oilseed rape were investigated. Data on the characteristics of the selected species formed the basis for modelling concepts. Deterministic, stochastic, population genetic and individual-based models or model concepts were developed and tested to map different aspects that may be important for gene drive applications.

The control of pest populations is one of the main objectives for possible applications of SPAGE in the olive fly, the most important pest in olive cultivation. Uncertainties exist with regard to the dispersal capacity of gene drive-bearing olive flies, high gene flow rates between different populations, and in particular with regard to the populationbottlenecks that occur regularly in winter, which can significantly increase or decrease genetic variability between subpopulations. The naturally occurring variability, in contrast to comparatively homogeneous laboratory conditions, leads to considerable and easily underestimated uncertainties about possible effects after releases.

The situation of the hazard and exposure potentials of oilseed rape, which arose from the allotetraploidisation of cabbage and turnip rape, is particularly complex. In oilseed rape it has been shown that crop plants can also form wild populations independently of cultivation and that transgenes persist in these populations. In addition, horizontal gene flow by hybridization is possible for several related species. It has been shown that there is an additional hybridisation network in the biological environment, which includes both rare and protected as well as widespread (weed) species. An advantage of a gene drive application in oilseed rape would be the reduction of transgene introgressions in wild populations or, more interesting from an economic point of view, the slowing down of the "pesticide treadmill". However, the application of an oilseed rape gene drive would be very costly, complicated to construct, very difficult to limit and practically impossible to supervise. Such an application should be avoided. In addition, a number of problems have been identified that may arise when other plant species are considered as potential target organisms for gene drives.

It is clearly foreseeable that if something goes wrong with the release of species carrying gene drives, corrective action is almost impossible. Limiting measures in the sense of the precautionary principle must take effect before release. For risk assessment, an additional step was proposed with the aim of evaluating the spatiotemporal controllability of gene drives. Taking into account (1) the natural biology of the targetorganisms, (2) their natural interactions with the environment and (3) the intended technical characteristics of the genetic modification, it should be possible to consider identifiable areas of "known unknowns" and uncertainties in the overall risk assessment and risk analysis. With regard to the possibilities and limitations of a stronger integration of the precautionary principleinto the governance of SPAGE, it could be shown in a first step that the environmental risk assessment (ERA) currently carried out in particular by the European Food Safety Authority (EFSA) does not sufficiently consider problems of dealing with complex systems characterized by nonlinearities and tipping points.

Although the current ERA guidelines already cover important aspects of hazard and exposure assessment, they also lack any precautionary aspects with regard to possible surprises (unknown unknowns). There is also a lack of guidelines for the weighting and assessment of the results of hazard and exposure analyses and, last but not least, a lack of clear consequences for precautionary risk management.

Action according to the precautionary principle is necessary if a particularly high damage potential is to be expected, but there is still non-knowledge or uncertainty regarding the exact nature and probability of occurrence of the threat. It is crucial that we cannot wait until everything is known or even the event has already occurred, because in this case corrective measures are no longer possible to a sufficient extent. On the other hand, precautionary measures cannot be justified solely by reference to a lack of knowledge. There must be comprehensible reasons for concern which indicate that an extensive hazard potential can be expected. Such concerns can relate to hazard and exposure potentials with regard to the technology used to intervene in systems as well as to weaknesses and tipping points within the affected systems. Precautionary measures should then not be limited to restrictions. Much more important is a precautionary design of the technologies and, if possible, a more resilient design of the affected systems. The key is to reduce the uncertainties arising from the effectiveness and scope of the respective technology, to reduce the depth and intensity of intervention, the power and scope of the technologies, and to develop technologies with a lower power and exposure potential. Self-limiting and retrievable genetic constructs could be first steps in this direction.

However, the precautionary principle must also be safeguarded by regulation. In the search for a viable solution, it is worth taking a look at the chemicals legislation under REACH. According to REACH the classification of certain properties of a substance as "reasons for concern" in the assessment of hazard and exposure potentials has a direct impact on precautionary risk management, namely a ban on its use with possible exceptions. In addition, the mere exposure to chemicals with very persistent and very bioaccumulative properties is recognised as being "of very high concern", even without a specific hazard hypothesis. These two steps in risk assessment and risk management could also integrate the precautionary principle into the regulation of new genetic engineering by classifying the extreme exposure potential, in particular of gene drives, as being of "very high concern" and leading directly to a prohibition of use with the possibility of derogations.

SPAGES and gene drives, this can be said in conclusion, threaten to reach tipping points on several levels (a) in (agricultural) ecosystems through an extreme expansion of exposure by organisms that have been deliberately released carrying gene drives, a process that enormously extends the scope for unexpected interactions, (b) through an increase in technological power and technical feasibility, which is now within the realm of the possible, and which should rather be approached with 'heuristics of fear' based on Jonas (1979) than with an unlimited form of overestimation of one's own capabilities, (c) in the risk governance of genetic engineering with the necessity of extended risk management and precautionary regulation and (d) with a view to social acceptance, without which no far-reaching innovations can be realised. Both socio-economic and (natural) ethical aspects play an important role here.

References

Frieß, J. L., von Gleich, A., & Giese, B. (2019). Gene drives as a new quality in GMO releases—a comparative technology characterization. *PeerJ, 7,* e6793. https://doi.org/10.7717/peerj.6793.

Jonas, H. (1979). Das Prinzip Verantwortung. Versuch einer Ethik für die technologische Zivilisation, Frankfurt/M.

Taleb, N. N. (2007). *The Black Swan: The impact of the highly improbable*. New York: Random House.

Permissions

All chapters in this book were first published by Springer; hereby published with permission under the Creative Commons Attribution License or equivalent. Every chapter published in this book has been scrutinized by our experts. Their significance has been extensively debated. The topics covered herein carry significant information for a comprehensive understanding. They may even be implemented as practical applications or may be referred to as a beginning point for further studies.

The contributors of this book come from diverse backgrounds, making this book a truly international effort. We would like to thank all the contributing authors for lending their expertise to make the book truly unique. They have played a crucial role in the development of this book. Without their invaluable contributions this book wouldn't have been possible. They have made vital efforts to compile up to date information on the varied aspects of this subject to make this book a valuable addition to the collection of many professionals and students.

This book was conceptualized with the vision of imparting up-to-date and integrated information in this field. To ensure the same, a matchless editorial board was set up. Every individual on the board went through rigorous rounds of assessment to prove their worth. After which they invested a large part of their time researching and compiling the most relevant data for our readers.

The editorial board has been involved in producing this book since its inception. They have spent rigorous hours researching and exploring the diverse topics which have resulted in the successful publishing of this book. They have passed on their knowledge of decades through this book. To expedite this challenging task, the publisher supported the team at every step. A small team of assistant editors was also appointed to further simplify the editing procedure and attain best results for the readers.

Apart from the editorial board, the designing team has also invested a significant amount of their time in understanding the subject and creating the most relevant covers. They scrutinized every image to scout for the most suitable representation of the subject and create an appropriate cover for the book.

The publishing team has been an ardent support to the editorial, designing and production team. Their endless efforts to recruit the best for this project, has resulted in the accomplishment of this book. They are a veteran in the field of academics and their pool of knowledge is as vast as their experience in printing. Their expertise and guidance has proved useful at every step. Their uncompromising quality standards have made this book an exceptional effort. Their encouragement from time to time has been an inspiration for everyone.

The publisher and the editorial board hope that this book will prove to be a valuable piece of knowledge for students, practitioners and scholars across the globe.

Index

www.ingramcontent.com/pod-product-compliance
Lightning Source LLC
LaVergne TN
LVHW051047220926
848485LV00001BA/39